Praise for **THE ART OF OPPORTUNITY**

"*The Art of Opportunity* is all about creatively discovering new growth opportunities for your company and crafting a collaborative strategy that will get you there. Not only is the content powerful, the design is stimulating for the eyes as well as the mind. Read this book and start innovating!"
—**Ken Blanchard,** Coauthor, *The New One Minute Manager®* and *Collaboration Begins with You*

"*The Art of Opportunity* will help trigger strategic renewal inside your organization. Creative, inspiring, fresh, and empirically grounded, this playbook to growth is bound to be an executive reference for many years to come."
—**Deryck J van Rensburg,** President, Coca-Cola Global Ventures

"From the coffee table to the conference table—*The Art of Opportunity* challenges our traditional business models and mind-sets while providing a path and approach to success."
—**Paul Snyder,** Vice President, Corporate Responsibility, InterContinental Hotels Group

"I love this book. Whether you're pushing an idea inside a large corporation, or creating your next big deal, this book gives you the framework, the tools, and the right questions to ask—not just to get you started—but to keep you going. In short, this book tells you how to organize your thinking with clarity, vision, precision and vigor."
—**Richard Black,** Chief Data Architect, Deutsche Bank; Former Chief Technology Officer, Bank of England

"Clear, artful, and inspiring. A highly readable, practical, step-by-step guide for anyone who wants to identify, design, and launch strategic growth initiatives."
—**Dave Gray,** Founder, XPLANE

"Innovation-obsessed visual thinkers, unite! This book may be your bible."
—**Sunni Brown,** Chief Human Potentialist; Best-Selling Author, *Gamestorming* and *The Doodle Revolution*

"*The Art of Opportunity* presents compelling evidence, based on extensive experience, for the application of business design thinking by combining strategic innovation with inspiration from others. Their pragmatic approach offers existing businesses a framework to innovate or adapt in complex changing environments. Loved it!"
—**Anne Bartlett-Bragg, PhD,** Managing Director, Ripple Effect Group

"*The Art of Opportunity* is not just a book, but a road map helping to guide companies as they venture down the innovation path in search of new growth opportunities. It offers readers an actionable lens that includes the vital components of storytelling and what it means to take a user-centered approach in the new world of business model design thinking."
—**Saul Kaplan,** Founder and Chief, Catalyst Business Innovation Factory; Author, *The Business Model Innovation Factory*

"*The Art of Opportunity* is not just another book on strategy! Sniukas, Lee, and Morasky created a practical tool to design future businesses. It is easy to read and use and spiced with brainstorming sessions you can run with your team. By using the book, I'm willing to bet companies can save fortunes on consultancy. Every business that wants to grow should go through this process."
—**Domenico Traverso,** President, Work Function Division, Danfoss Power Solutions

"Visually beautiful, *The Art of Opportunity* is a fresh, new take on design thinking that offers practical tools for strategic innovation."
—**Mark Polson,** Vice President Creativity and Strategic Capability Building, Estée Lauder Companies

"This book will dramatically alter both your business and personal life as you experience the very useful approach to innovation and strategy. Get on board and enjoy this inspirational journey."
—**Dr. Bob Lorber,** CEO, The Lorber Kamai Consulting Group; Coauthor, *Putting The One Minute Manager to Work* and *Doing What Matters*

"*The Art of Opportunity* is not your typical strategy book. It offers a fresh approach to designing growth strategy with methods that are equally valuable to leaders of established businesses and new ventures."
—**Jeff Wright,** Vice President, Strategy and Marketing, Autodesk

"Opportunities abound. We confront them every day, but they rarely appear as opportunities and do not come with labels to identify them as opportunities. As a blind person myself, I'm excited that the authors of this book use visualization as a tool to bring opportunities into our businesses and lives. Read, learn, and imagine."
—**Jim Stovall,** Best-Selling Author, *The Ultimate Gift*

"*The Art of Opportunity* is a concise, richly detailed primer for organizations seeking growth through innovation, and change through disruption."

—Jay Samit, Best-Selling Author, *Disrupt You: Master Personal Transformation, Seize Opportunity,* and *Thrive in the Era of Endless Innovation*

"Refreshingly pragmatic advice for business growth and development. This visualization approach was successful in aligning resources to solve my complex business issue."

—David Lary, Vice President, Commercial Channel Partner Development and Programs, Print and Personal Systems, Hewlett-Packard

"*The Art of Opportunity* humanizes business innovation in its approach and accessible, visual presentation. A must-read for entrepreneurs who aim to disrupt industries."

—Chip Joyce, CEO & Cofounder, Allied Talent

"In my job I am expected to know this information, but this book presents all necessary information so clearly, it's a piece of art! I am convinced that *The Art of Opportunity* will help thousands of managers gain knowledge much quicker than they could ever have done so in their working life. Well done, and thank you so much for the effort!"

—Silvester de Keijzer, Board Member, Swiss Made Im.

"*The Art of Opportunity* presents a clear field guide for creating new value for organizations. The methods and stories affirm that the most valuable parts of business are more art than science. Ignore at your own risk."

—Michael Graber, Managing Partner, Southern Growth Studio

"A wonderful new tool for those feeling 'stuck' with old approaches and business processes that are no longer effective."

—Max Thelen, Former CEO, Filter

"Proven strategies from market innovators? Check. Step-by-step guides to structure your exploration? Check. Innovative visual explanations to engage and inspire? Triple-Check. If you're a market-maker, don't miss this book."

—Kevin Tate, Cofounder, StepChange Group

"*The Art of Opportunity* is an innovation cocktail with a nice twist of visual thinking that will inspire application and growth."

—Jim Wallace, Global Head of Agency Strategy and Management, Hewlett Packard

"What a bold and compelling book! It offers a thought-provoking new perspective on strategy and some important food for thought on how to discover and seize new opportunities. The strengths of *The Art of Opportunity* lie in its practical approach to validate strategies. It does not just claim that experimentation is important but shows how you can actually validate a strategy through experimentation. Bravo!"

—Anja Förster and Peter Kreuz, Best-Selling Authors, *Entrepreneurs and Angel Investors*

"A fresh approach to business design thinking that should give executives, decision makers, and leaders of strategic initiatives plenty of ideas for improving outcomes and delivering value."

—Aaron Smith, Editor, ProjectsAtWork.com

"*The Art of Opportunity* is a beautifully crafted piece of work, so richly illustrated that at times I felt like I was reading a graphic novel. It is packed with fresh and varied case studies, most of which I hadn't come across before. But more importantly, this book is really useful. Whether you're an entrepreneur in the first throes of wrestling with a business idea, or a seasoned strategist at a large organization looking for new paths toward new opportunities, this is one of those rare how-to guides that actually overdelivers on the how-tos. My own copy will be well thumbed, for sure."

—Mark Barden, Coauthor, *A Beautiful Constraint*

"Strategy made fun! An innovative way of making strategic innovation come alive. This book will bring out the best of both the right and left half of your brain. Great for novice strategists and old hands alike."

—Ron Meyer, Professor, Corporate Strategy Tias Business School, Tilburg University

"Growing a business is all about identifying and leveraging opportunities. But where to start? How to know where to look? This book not only makes you change your thinking, it gets you moving, too, with novel and practical exercises that provide managers with a customizable pathway to growth."

—Professor Helen Perks, University of Nottingham, UK

"How to make something challenging and complex easier to understand [and], easier to do by breaking it down into manageable and clear tasks. This book gives you very practical steps and advice on building new initiatives."

—Stuart Curley, Vice President Enterprise Architecture, Northgate IS

"For anyone trying to navigate the complexities of value creation, *The Art of Opportunity* gets to the heart of how to operationalize entrepreneurship and innovation as the real engines behind strategic growth."

—Wayne A. Simmons and Keary L. Crawford, Authors, *GrowthThinking: Building the New Growth Enterprise*

"*The Art of Opportunity* provides a refreshing view on how to grow performance in an increasingly visual and collaborative world. It is compelling and simple and just what the business landscape needs. Worthy read."

—Crystal Fernando, Head of Global Commercial Delivery, InterContinental Hotels Group

"A fantastic and beautiful design-thinking manifesto for the visual learner in all of us."

—Matt Moog, CEO, PowerReviews

"Sniukas, Lee, and Morasky have crafted a beautiful, full-color journey to innovation that takes you from beginning to end. The engaging visuals draw you in to understand the approach, while the simple step-by-step instructions get the thoughts out of your head and ready to execute. A must-read for anyone looking to tackle an opportunity they have spotted."

—C. Todd Lombardo, Chief Design Strategist, Fresh Tilled Soil; Coauthor, *Design Sprint*

"Step by step, *The Art of the Opportunity* encourages a fresh, visual look at how to deliver new value to customers."

—Doug Van Aman, Principal, Van Aman Communications

"Sniukas, Lee, and Morasky have artfully made complicated strategic design principles accessible to anyone looking to grow or build a business. *The Art of Opportunity* is a visually engaging guidebook that helps businesses of any size tap into their creativity to develop effective business strategies."

—Catherine Palmer, Senior Industry Marketing Manager, Autodesk, LLC.

"Whilst managing Apple Education in EMEA, we discovered the power and resilience of visual communications, giving us the ability to translate complex digital engagement into simple, informative and meaningful illustrations and graphics. Today, *The Art of Opportunity* helps move this work forward another important stage."

—Alan Greenberg, Former Apple Director Education EMEA Currently Advisory Board of 8GT Fund

"*The Art of Opportunity* is a key innovation masterpiece that is going to help me with our work with intrapreneurs at the largest entertainment media companies in the world."

—John Huffman IV, Founder, HUFFMAN CO

"*The Art of Opportunity* blazes a welcome trail to successful business growth."

—Kevin O'Keefe, Business leader and bestselling author, *The Average American*

"Impactful and practical. An excellent guide for strategic reflection and action."

—Omar Baig, Head of Digital, Knowledge and Information Service, OECD

"Being, Knowing, and Doing are must-have disciplines for any team to succeed. This book helps guide teams with crisp insights and methodologies to rapidly frame innovation projects."

—Ed Soo Hoo, Industry Fellow, UC Berkeley Center for Entrepreneurship and Technology

"*The Art of Opportunity* clearly details how to achieve success through proven strategic innovation and visual thinking methodologies. The authors expertly plot a path along the journey based on years of experience in helping companies realize significant growth. Let great things happen to you by reading this book."

—K.C. Teis, Vice President of Experience Design, Rackspace

"In working with over 150 SMEs in the last 12 months, I have found they want business support that is practical, visual, and guides them to make the right decisions by inspiration and example. *The Art of Opportunity* does this brilliantly. It brings key visual tools into one place and shows you how to use and implement them at every stage for your business' success. Excellent book, a must buy."

—Mark Copsey, Director, RedKite Innovations; Growth coach and lecturer, Leeds Beckett University

"A clear and practical approach to uncovering growth opportunities hidden within your business and how to make them a reality and driving the top line."

—Robert McKinnon, Former SVP and Head of PMO, InterContinental Hotels Group

"Destined to become 'the bible' of incorporating business design thinking into building growth. It's a must-read."

—Lars Crama, Chief Commercial Officer, InnoLeaps

"An easy-to-follow blueprint for customer-centric innovation to drive business growth. Had I had this book 10 years ago, I'd have saved tons on agency fees!"

—**Heinz Waelchli,** Chief Customer Officer at SnapAV

"It's not often a series of artful lateral choices are presented so elegantly that they can be pragmatically implemented like science. A must-read for leaders looking to innovate with their existing teams to create new value."

—**Shayne Smart,** Founder, Geneva Conventions in Pictures

"Read this book with a pen in hand, because you'll want to start creating your next strategic innovation straight away. This practical guidebook to business design thinking will help you make the most of your customer opportunities."

—**Simon Terry,** Chairman, Change Agents Worldwide

"Practical. Transformative. Actionable. Leaders should equip their staff with this innovative guide. Follow the instructions. Your business will grow."

—**Michael Neil,** Former Director, Digital Marketing, Franklin Square Capital Partners

"*The Art of Opportunity* brings you the artful innovation approach to helping you discover new growth opportunities that you would expect given the title. The book contains many useful tools including some nice approaches to customer journey mapping and practical tools for workshop leaders. It also does a good job of making design thinking and growth planning accessible for the nonexpert innovator. Keep innovating!"

—**Braden Kelley,** Cofounder, InnovationExcellence.com, Author, *Charting Change and Stoking Your Innovation Bonfire*

"The question of how to effectively develop innovative business models with linkage to strategy innovation has remained unsolved—both in theory and in practice. This is one of the few books that unites the three components of strategy, business model innovations, as well as creativity techniques, by providing useful methods for corporate entrepreneurs."

—**Dr. Daniel Liedtke,** COO, Hirslanden Private Hospital Group

"Sniukas, Lee, and Morasky provide a powerful framework that encourages and empowers readers to apply their own creativity to developing their growth strategy."

—**Robert Shepherd,** Chief Development Officer and SVP Development, Design and Openings, Europe, InterContinental Hotels Group

THE ART OF
OPPORTUNITY

HOW TO BUILD GROWTH AND VENTURES THROUGH STRATEGIC INNOVATION AND VISUAL THINKING

WILEY

Cover image: Matt Morasky

This book is printed on acid-free paper.

Published by John Wiley & Sons, Inc., Hoboken, New Jersey.
Published simultaneously in Canada.

Design: Bryan Zentz, Matt Morasky.
Illustration: Matt Morasky, Nathan Stang, Guonan, Elisa Gipson, Susanne LeBlanc, Graham Barey.

Library of Congress Cataloging-in-Publication Data:

Names: Sniukas, Marc, author. | Lee, Parker, author. | Morasky, Matt, author.
Title: The art of opportunity : how to build growth and ventures through strategic innovation and visual thinking / by Marc Sniukas, Parker Lee, Matt Morasky.
Description: Hoboken : Wiley, 2016. | Includes index.
Identifiers: LCCN 2016002292 | ISBN 978-1-119-15158-6 (pbk.) | ISBN 978-1-119-15160-9 (ePDF) | ISBN 978-1-119-15159-3 (ePub)
Subjects: LCSH: Business planning. | Technological innovations—Management. | Organizational change.
Classification: LCC HD30.28 .S615 2016 | DDC 658.4/012—dc23 LC record available at http://lccn.loc.gov/2016002292

Printed in the United States of America
10 9 8 7 6 5 4 3 2 1

THE ART OF OPPORTUNITY

HOW TO BUILD GROWTH AND VENTURES THROUGH STRATEGIC INNOVATION AND VISUAL THINKING

MARC SNIUKAS PARKER LEE MATT MORASKY

WILEY

To my wife for being patient with me and encouraging me to persist.

—Marc Sniukas

To my RFL and my family for unconditional love and support.
And to Aric, for helping me to fly.

—Parker Lee

To Bob and Sandy, whose support has never wavered, not once.

—Matt Morasky

CONTENTS

1 ARTFUL INNOVATION

2 DISCOVER YOUR NEW GROWTH OPPORTUNITY

3 CRAFT YOUR STRATEGY

4 LAUNCH YOUR NEW GROWTH BUSINESS

5 MASTERING THE ART: BUSINESS DESIGN THINKING

FOREWORD

When many of today's leaders joined the workforce, "innovation" was synonymous with research and development or process efficiencies—the hallmarks of traditional competitive advantage. Little did any of us know then, that in our lifetimes an entire occupational discipline would emerge to keep companies "innovative" or continuously inventive. Or that titles like "director of innovation" and "chief innovation officer" or even "chief imagination officer" would decorate corporate organization charts.

But it did. And for good reason. The relatively short span of time in which we've seen some of the titans of industry displaced by "innovative" start-ups put the entire business world on notice. And the message is clear: merely maintaining your position is no longer sufficient. New growth, the kind associated with genuine innovation, that will bring value to your customers, your business, and even the world around you is the only way to ensure survival.

This urgency is not only felt by those with the word "innovation" etched into their job titles, but from the top of the company to the bottom, we are now all corporate innovators. With all these people focused on the problem, we should have it solved, right? Well . . . , no.

The problem is, finding and capitalizing on new growth opportunities is hard—especially for established organizations that are often hampered by outdated mindsets, legacy business models, or large-scale bureaucracies. Core competencies can morph into corporate rigidities if we're not strategically alert and careful. Under these types of circumstances, the ability to "think outside the box" and create new growth initiatives is difficult. But with increased urgency comes the need to find a new path to growth—one that isn't rocket science. What we need is a road map to help discover, catalyze, and curate opportunities to deliver real growth. That's what this book helps us do.

Marc, Parker, and Matt have authored a rich, compelling journey for anyone pursuing new growth. In the process of introducing new ways of thinking about growth strategy and strategic innovation such as noncustomers, essential customer needs, and the building blocks of business and revenue models, they also introduce us to new ways of working. Throughout the book visual thinking methodologies and other business design thinking principles improve how you approach crafting and executing the strategy to build your new growth business. And these are supported by visualizations, tools, and templates that help you apply the approach to your own needs.

Additionally, they comfortably switch from well-known examples that illustrate concepts to fresh, unpublished case studies that serve to inspire the reader. In the end, they have created a provocative playbook grounded in rigorous academic research combined with practical, immersive experience.

While "innovation" isn't new, *The Art of Opportunity* makes it much more accessible to everyone. And that's something every corporate entrepreneur searching for new growth urgently needs.

–Deryck J van Rensburg, President, Coca-Cola Global Ventures

ACKNOWLEDGMENTS

We offer our sincere thanks to our colleagues from who provided insight and expertise that was instrumental in researching, designing, and writing *The Art of Opportunity.*

We are deeply grateful to the brilliant design and production work of Bryan Zentz, who went over and above the call of duty to create a book that is an enticing work of art.

Kudos to our stalwart marketing team:
Tori Dunlap, Lucy Kelly, Hannah Vogel, and Heidi Hinshaw for assistance with marketing.

For providing design, illustration, and production talent, we applaud:
Graham Barey, Hermen Lutje Berenbroek, Elisa Gipson, Guonan, Susanne LeBlanc, and Nathan Stang.

A special thanks to our XPLANE colleagues, both past and present:
As a source of inspiration, support, encouragement, and professional camaraderie.

For generous use of office space: Jon McDonald at The Good and David Howitt of The Meriwether Group.

We are also thankful for the encouragement, support, and talent of:
John Willig of Literary Services, Inc.; Lia Ottaviano, Peter Knox, Lauren Freestone, and Matt Holt of John Wiley & Sons; Joan Bosisio, Jen Heady, and Jacqueline Bole of Stern Strategy Group for strategy and public relations support; Max Thelen for legal counsel, advice, and general wisdom; Guanon for website design; Ann Smith, Lisa Lavora-De Beule, and Natalie Hanson of A.wordsmith.

For sharing their pearls of wisdom with us, we are immensely grateful to:
Bob Kelly, Microsoft; Anne Bartlett-Bragg, Ripple Effect Group; Mark Baker, Oracle; Sunni Brown, SB Inc; Tom Williams, Salesforce; Jim Kreller, Vanco Payment Solutions; Geoffrey Moore, Chasm Institute; Loius Patler, The B.I.T. Group; Nancy Duarte, Duarte Design; Dave Gray, XPLANE; Robert L. Lorber, PhD, Lorber Kamai Consulting Group; David Allen, David Allen Company; Jay Samit, Robert I. Sutton, and Mark Polson, Estée Lauder Companies; Paul Snyder, IHG; Robert McKinnon; Michael Graham; James Macanufo; Chip Joyce, Allied Talent; Jim Wallace, HP; Jeff Wright, Autodesk; Ron Diorio, The Economist; Soon Yu, VF Corporation; Omar Baig, OECD; Richard Black, Deutsche Bank; Bryson Koehler, Weather.com; Stuart Curley, Arriva; Gary Fisher, UC Davis; Patrick van der Pijl, Busines Models Inc.; Mark Barden, eatbigfish; Glenn A. Oclassen, Jr., Replicon; Penny Fondy, Wits' End Productions; Michael Graber, Southern Growth Studio; Dion Hinchcliffe, 7Summits; Jonas Koffler and David Roberts, The Difference; and our research participants, especially Dr. Daniel Liedtke and Klinik Hirslanden.

And, finally, many thanks to our community of Art of Opportunists for volunteering to review early content and provide their wisdom, advice, and support:
Apostolos Apostolides, Patrick Attallah, Omar Baig, Anne Bartlett-Bragg, Tanner Bechtel, Richard Black, Jacqueline Bole, Sunni Brown, Gary Burt, Mike Colarossi, Mercedes Concepcion-Gray, Mark Copsey, John Cousineau, Lars Crama, Stuart Curley, Hilary Curry, Kord Davis, Silvester de Keijzer, Peter de Wit, Matthew Dennison, Alexander Doujak, Crystal Fernando, Claire Francois Michael Graber, Dave Gray, Alan Greenberg, Kamal Hassan, Jen Heady, Kira Higgs, Heidi Hinshaw, Bob Kelly, Lori Kenney, Timur Khojayev, George Kurtyka, David Lary, C Todd Lombardo, Liz Lufkin, Smita Master, Rob McKinnon, Horacio Miranda, Karl Mundorff, Martina Naerr-Fuchs, Eva Nell, Kate Niederhoffer, Niall O'Doherty, Andres Pacheco, Vincenzo Pallotta, Catherine Palmer, Louis Patler, Gunter Pfau, Max Rehkopf, Stuart Rolinson, Fernando Saenz-Marrero, Brendan Sherry, Paul Snyder, Jodi Sweetman, Kevin Tate, The Doujak Team, K. C. Teis, Max Thelen, Rafal Tomczyk, Mike Turner, Heinz Waelchli, Jim Wallace, Margaret Ward, Terry Ward, Scott Wilkinson, and Thomas Williams.

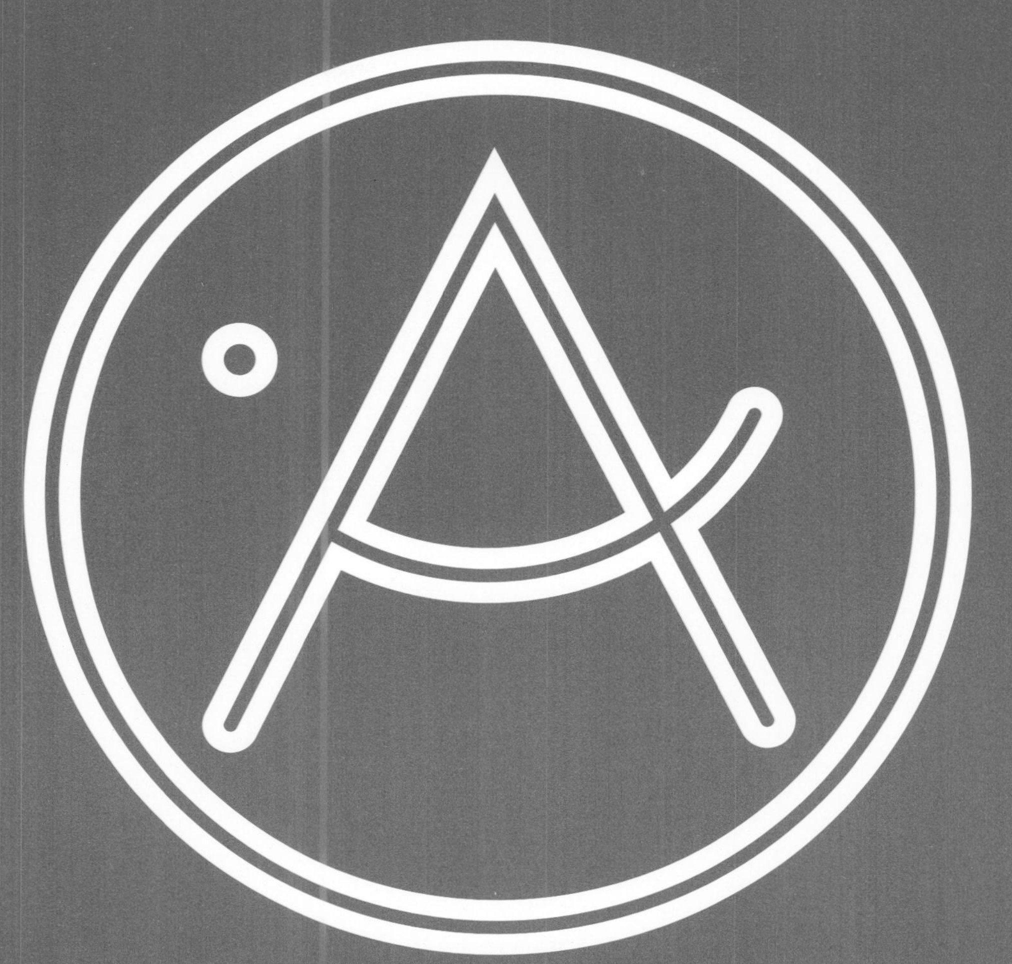

DO. OR DO NOT. THERE IS NO TRY.

—YODA *Jedi Master*

We wrote this book to answer the question: How can established companies create new growth strategies and businesses from within their organizations?

By looking at how successful companies addressed this question and overcame their growth challenges, we envisioned and framed an approach that reduces risk, delivers results faster, and has a higher likelihood of success. But our approach should not be misconstrued as a guaranteed method. Finding and seizing new growth opportunities is an art. And like any art, it requires personal dedication, professional rigor, and a passionate drive to succeed. Our experience has shown us that those who apply the strategic and visual thinking methods described in this book not only improve their chances of discovering growth opportunities, but also are in a better position to realize the kind of success experienced by today's most innovative companies.

In writing and designing *The Art of Opportunity*, we have applied many of the same processes and principles presented in the book to our own work. Our collaborative process was greatly accelerated by the same type of visual thinking activities and methods for identifying, designing, and launching opportunities. We have also applied a diverse team-based approach, with each author bringing a unique set of professional and cultural experiences to broaden our perspective not only on the subject matter, but also on the reader's journey. Finally, we have followed a system of active iteration, creating the book in cycles and seeking input whenever possible to improve the content and presentation.

Our thought has been inspired by the works of Gary Hamel, Clayton Christensen, W. Chan Kim, Renée Mauborgne, David Teece, Dave Gray, Don Kohlberg, Jim Bagnell, and David Kelley, among others, and is informed by our professional work with companies around the globe. The concepts we outline in this book were cultivated over the last 20 years out of our own original academic research and our experience applying what we learned and developed to help organizations grow, innovate, and transform.

THE ART OF OPPORTUNITY READER'S JOURNEY

To help the reader better understand and, ultimately, craft new growth opportunities, *The Art of Opportunity* has been structured around the idea of a "Reader's Journey." This journey introduces concepts, demonstrates principles, and presents activities in a way that allows the reader to both understand and practice the art of opportunity. While we recognize that there are virtually an unlimited number of routes to find new growth, we hope the lessons learned along the way will enable the reader to more successfully explore and create their own growth journey. The "Reader's Journey" includes:

1. **Core concepts:** with supporting illustrations and diagrams.
2. **Inspirations:** examples of how other organizations have put the concepts into practice to generate breakthrough growth.
3. **Sparks:** tactics that encourage the reader to reflect on the concepts in relation to their own growth journey.
4. **Activities:** "Learn-by-doing" workshop-based exercises designed to guide your team through the growth process.
5. **Templates:** simple visual formats (found at the end of each section) for assembling and presenting key findings generated during your growth journey.

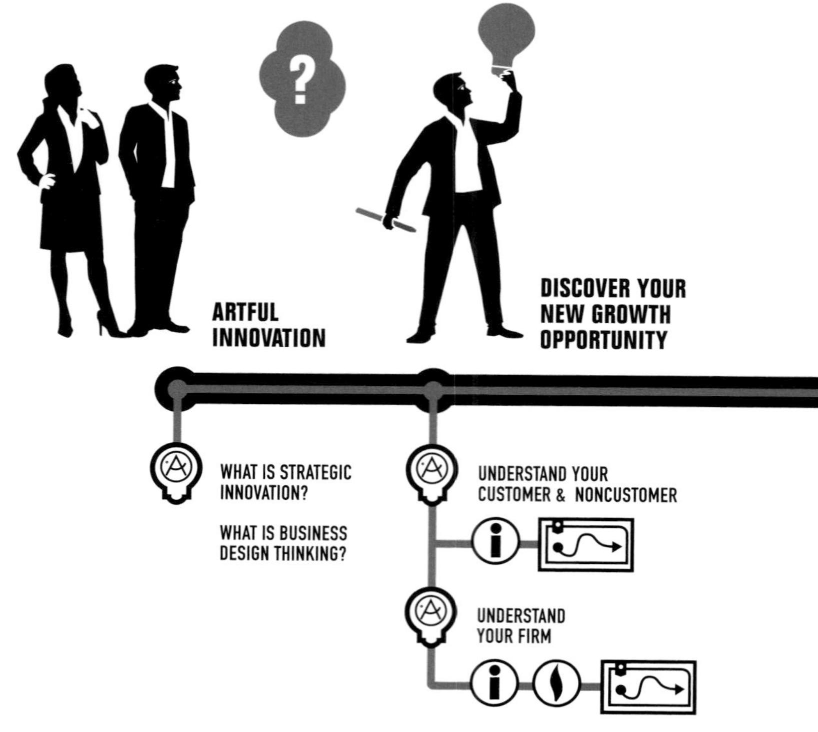

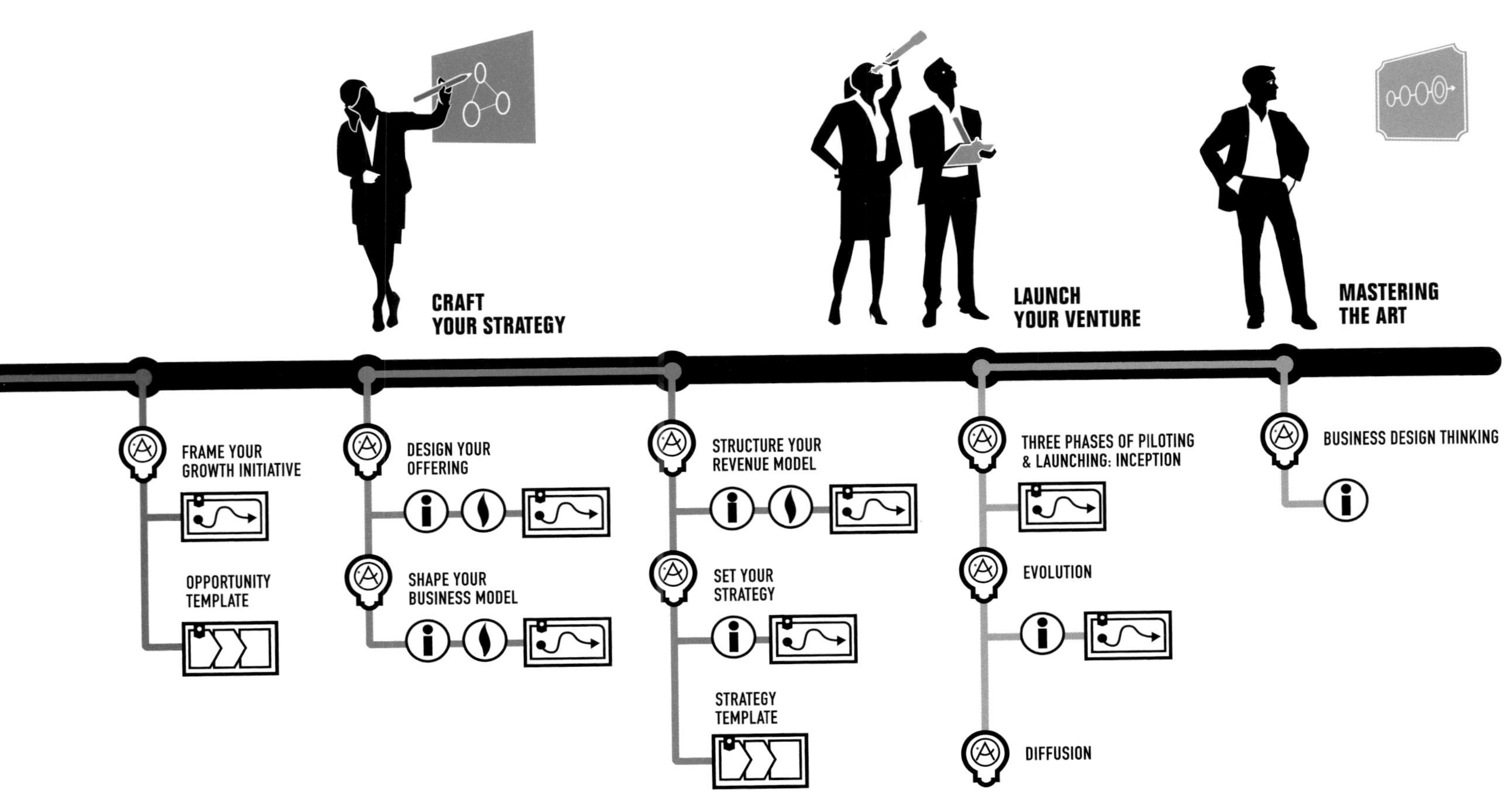
CRAFT YOUR STRATEGY
LAUNCH YOUR VENTURE
MASTERING THE ART
FRAME YOUR GROWTH INITIATIVE
OPPORTUNITY TEMPLATE
DESIGN YOUR OFFERING
SHAPE YOUR BUSINESS MODEL
STRUCTURE YOUR REVENUE MODEL
SET YOUR STRATEGY
STRATEGY TEMPLATE
THREE PHASES OF PILOTING & LAUNCHING: INCEPTION
EVOLUTION
DIFFUSION
BUSINESS DESIGN THINKING

INNOVATION

DRIVES US.

ARTFUL INNOVATION

WE SEIZE GROWTH OPPORTUNITIES IN A TARGETED WAY AND BENEFIT FROM THE DYNAMIC DEVELOPMENT OF NEW BUSINESS AREAS IN THE LONG TERM.

—**PROSIEBENSAT.1 MEDIA** *Annual Report 2014*[1]

CHAPTER 1

When Thomas Ebeling became the new CEO of ProSiebenSat.1 Media AG (Pro7) in 2009, he faced a familiar problem: how to grow the business. Pro7 was the leading media company in the German TV advertising market with their core business of free TV—financed by advertising. Mr. Ebeling's challenge was how to grow the business and achieve over €1 billion in incremental revenues by 2018. Resorting to traditional strategic moves, expanding existing businesses and offerings, would certainly not do the trick.

Pro7 crafted a unique strategy for a new business area making TV advertising available to start-ups, and small and medium-sized companies, two customer segments that traditionally couldn't afford TV advertising and were seen as unprofitable by the industry. One year after the new business had been launched, it had already generated €20 million in profit. Five years later, Pro7 had achieved tremendous success with its new strategy. In fact, the strategy worked so well that in 2015, the €1 billion plus target was increased to €1.85 billion.

Business professionals and scholars admire achievements like these and wonder how Pro7 tapped into new markets and developed new business areas. Finding and seizing opportunities for new growth is, after all, the Holy Grail of business. And if Pro7 didn't use traditional growth methods, exactly how did they arrive at new offerings, complete with innovative business models and revenue models, to reach entirely new customer segments?

WHAT IS STRATEGIC INNOVATION?

THE ART OF OPPORTUNITY is about how strategic innovation and business design thinking can grow existing businesses and create completely new ones by discovering opportunities for new growth and crafting strategies to seize these opportunities. Strategy is essentially about making choices about where to play and how to win. Strategic management theories offer frameworks to guide our thinking, help develop answers to these questions, and make such choices. The approach and concepts of *The Art of Opportunity* provide fresh and modern ways to look at these questions, enabling you to come up with more innovative answers than yesterday's traditional strategic management approaches might offer.

TRADITIONAL STRATEGIC MANAGEMENT

TRADITIONAL STRATEGY

i. Where to play
ii. How to win

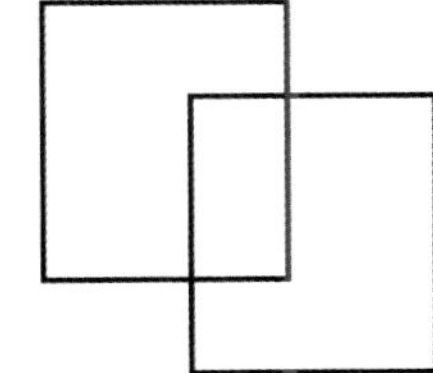

VS.

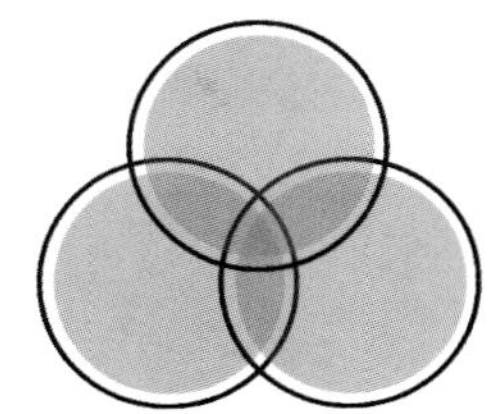

STRATEGIC INNOVATION

i. Where to play
ii. How to play
iii. How to win

ON A BUSINESS LEVEL, traditional strategic management[2] is primarily concerned with where to play and how to win. Where to play is framed as a choice of industry and product/market combination. Simply speaking, you pick an industry, say air transportation, and a market, for example continental flights in the United States, and you define your offering within this market, for example low-cost direct city-to-city flights.

How to win is mostly defined as achieving competitive advantage. Michael Porter's classic generic choices[3] about how to achieve competitive advantage are to either (1) be a cost leader, (2) differentiate your offering, or (3) focus on a niche. Treacy and Wiersema[4] offer three choices to win: (1) product leadership (offer the best product by focusing on product innovation); (2) operational excellence (be a price and convenience leader by focusing on low cost, lean and fast production, and speedy delivery); or (3) customer intimacy (win by creating loyal customers through tailored offerings and focusing on customer relationships).

Strategy development and execution thereby follow a linear process of analyzing the situation and environment, followed by developing a strategy, and finally executing it. The underlying principle is that the development of the strategy has to be completed before the strategy can be executed.

To be clear: we don't suggest that these traditional strategic management approaches do not work. For some organizations and in certain industries, they work extremely well, if applied in the right way. Yet a lot of companies also struggle when attempting to achieve their growth and innovation targets with these traditional frameworks.

STRATEGIC INNOVATION

HOW DOES OUR TAKE on strategic innovation differ from these more traditional approaches?

First of all, we shift the objective from focusing on achieving competitive advantage by simply being cheaper or different to finding and seizing opportunities by creating value. Traditional strategy is focused on the company, trying to position the company as being a cost leader, being different, focusing on a niche, or something similar, as we have seen. But being cheaper or different alone is simply no longer enough to be successful (if it ever was). *The Art of Opportunity* takes an entrepreneurial stance, looking beyond positioning your company to a larger holistic perspective that involves creating value for your customer, your firm, and your business ecosystem. Only by creating value for a multitude of stakeholders does your company have the potential to be successful. And creating value is achieved through more than simply offering a cheap or different product, to include products, services, the entire customer experience (CX), your business model, and your revenue model.

Having described the differences, let's examine *The Art of Opportunity's* approach to designing your strategy. The book offers a fresh perspective to look at three areas: (1) Where to play, (2) How to play, and (3) How to win.

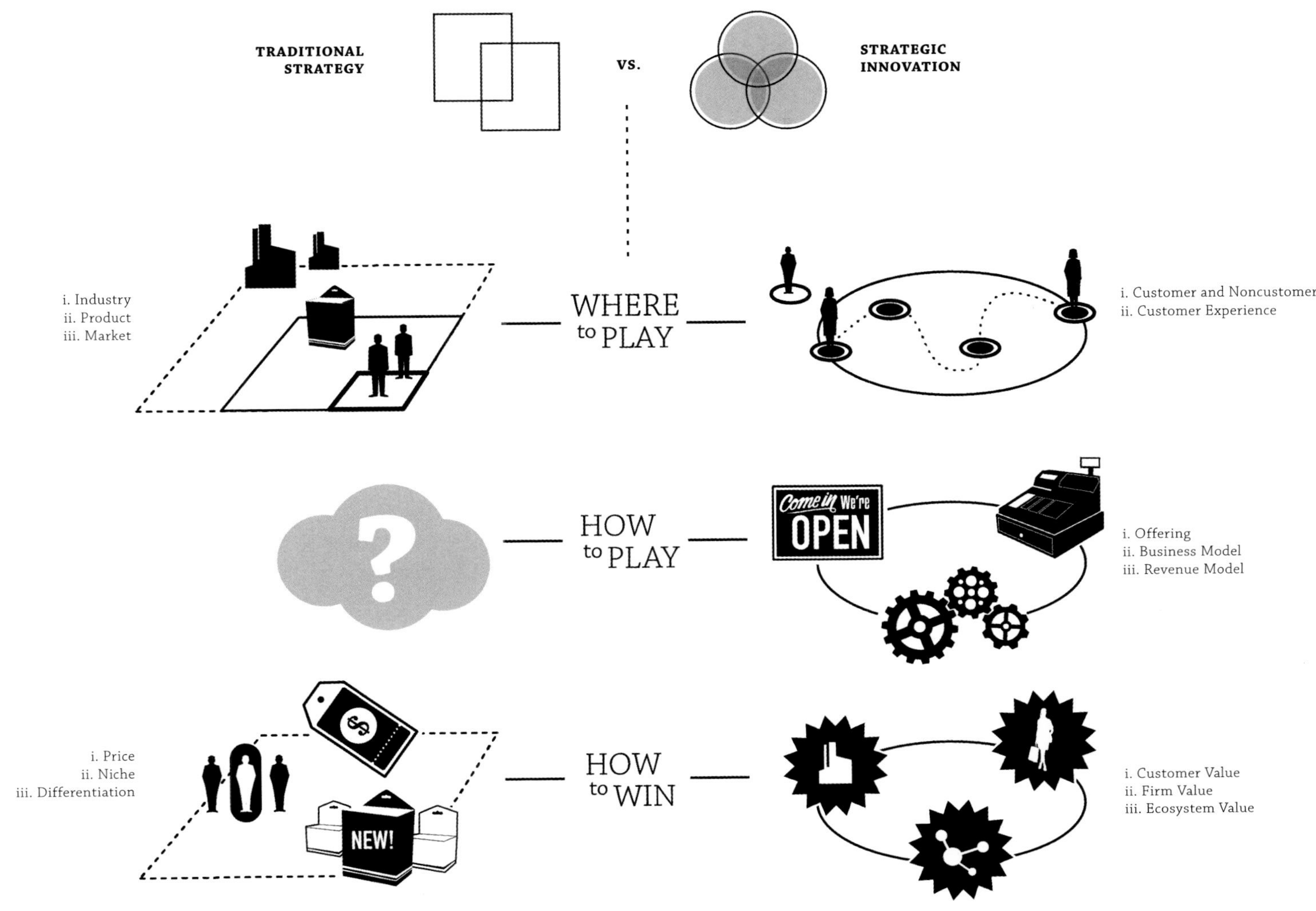
TRADITIONAL STRATEGY
VS.
STRATEGIC INNOVATION
i. Industry
ii. Product
iii. Market
WHERE to PLAY
i. Customer and Noncustomer
ii. Customer Experience
HOW to PLAY
Come in We're OPEN
i. Offering
ii. Business Model
iii. Revenue Model
i. Price
ii. Niche
iii. Differentiation
NEW!
HOW to WIN
i. Customer Value
ii. Firm Value
iii. Ecosystem Value

STRATEGIC INNOVATION

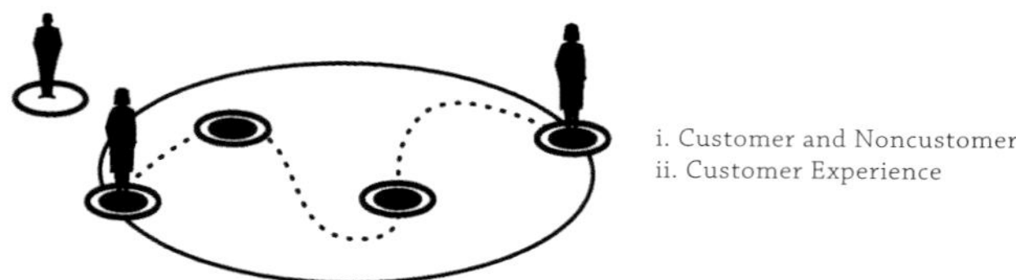

WHERE TO PLAY: *Find Your Opportunity*

Where to play is all about finding your new growth opportunities. Instead of focusing on industries, markets, and competitors, we focus on exploring:

- **Customers** and noncustomers,
- **Customer needs** and expectations,
- The **customer experience**, including barriers to consumption and hurdles to satisfaction.

Opportunities are a function of the chosen customer segment, its needs, and expectations toward the solution offering, and current barriers to consumption or hurdles to a satisfactory customer experience.

HOW TO PLAY: *Craft Your Strategy*

How to play is all about crafting the strategy and designing the business required to seize these opportunities. Defining how to play requires you to craft your:

- **Offering:** The unique blend of products, services, and the customer experience.
- **Business model:** The necessary set of activities to create and deliver your offering, in a specified sequence of these activities, employing the skills, capabilities, and assets necessary to do so and identifying who provides them, plus how you work with your partners.
- **Revenue model:** The combination of your revenue streams, pricing mechanisms, and payment schemes.

i. Customer Value
ii. Firm Value
iii. Ecosystem Value

HOW TO WIN: *Create Value*

Instead of simply addressing cost and pricing or product/service differentiation, we focus on creating:

- **Customer value:** Solving your customers' needs better than anybody else by removing barriers to consumption and hurdles to satisfaction.
- **Firm value:** Crafting a strategy that will generate value for your company in terms of opening up further opportunities and operational and financial benefits.
- **Ecosystem value:** Creating strategic, operational, and financial value for your partners and the larger ecosystem your company is embedded in and relies on.

Finally, we show how the process for strategy making and execution and for building new growth businesses is neither entirely linear nor completely iterative. Instead we will illustrate how companies go through an iterative process consisting of three phases characterized by a bias toward action over analysis and planning.

WHAT IS BUSINESS DESIGN THINKING?

IF STRATEGIC INNOVATION FOCUSES on the content of your new growth strategy and the process of crafting that strategy, business design thinking focuses on the practices that enable your team to achieve success more effectively and efficiently.

In short, business design thinking is a collection of principles (of which visual thinking is a key methodology) to help understand, address, and develop solutions to business problems. It can also be considered a strategic mind-set (or way of working) that focuses on understanding audiences, visualizing ideas and information, working collaboratively, and learning iteratively, all while keeping an eye on a holistic picture. This approach has been proven to open new channels to creativity, actively engage participants and stakeholders, build clarity and consensus, and accelerate speed to market.

THE FIVE PRINCIPLES OF BUSINESS DESIGN THINKING:

1. **Keep a human-centered focus:** Put people, not objects, at the heart of your story. An empathetic, human-centered focus creates value for not only the customer, but all stakeholders, including employees, shareholders, suppliers, and vendors.
2. **Think visually and tell stories:** Visualization enables us to more easily and clearly share our ideas and develop them with others. Visual storytelling brings ideas to life and creates the understanding and alignment that accelerates decision making.
3. **Work and co-create collaboratively:** Bring together diverse perspectives. Creating solutions to a shared problem within a multi-disciplinary group builds support and can generate breakthrough ideas.
4. **Evolve through active iteration:** Build to understand. Iteration enables you to learn reflectively during the process

Human-Centered Focus

Visual Thinking

Co-Creation

Iteration

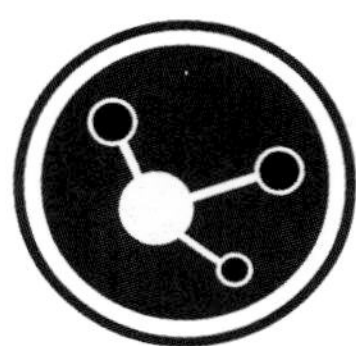
Holistic Perspective

of creation. The result is quicker, more successful adaptation and evolution of your ideas, solutions, and/or strategies.

5. **Maintain a holistic perspective:** View the organization as a dynamic, open system of interrelated processes. This vantage point can help you identify opportunities to break down silos, improve efficiency, and create context for understanding.

The Art of Opportunity incorporates each of the five principles of business design thinking to embody the practitioner's way of working. Most prominently we have incorporated visual thinking—from illustrations that accelerate understanding of new ideas to visually based activities and templates to drive clarity, spur collaboration, and build support. The process encourages cyclical development that supports iterative learning and we actively promote identification of value through holistic perspectives.

Executives applying the business design thinking way of working will develop capabilities and practices different from those of their peers. We admit we're biased. We have seen how businesses can apply business design thinking to create the greatest value for their customers, employees, and ecosystems. Employees feel empowered to use their judgment to make decisions. They are trusted and supported to do the right thing and rewarded for their creativity and initiative. Customers appreciate the resulting quality and value provided by the firm's products and services, thereby increasing the firm's financial and brand value. Partners within the ecosystem gain value similarly, through increased business and by association with the firm.

GETTING STARTED

OUR RESEARCH HAS FOUND that growth initiatives have a greater chance of success when they share the following common characteristics:

- A dedicated, diverse team.
- Visible and acknowledged sponsorship of, and commitment to, the initiative from the leaders of the organization.
- Dedicated time, resources, funding, and physical space.
- Clear goals and expectations, and time-bound parameters.
- Open, continuous communication—to everyone, all the time.
- Employ visual thinking and storytelling to manage complexity.

Of these characteristics, the most critical is a dedicated, diverse team. You will need a mix of interdisciplinary thinkers representing key stakeholder groups to execute the strategic innovation process. Seek individuals who can devote (or have been given leave to dedicate) the required time and possess the following business design thinking skills, characteristics, and capabilities:

- Applies an open mind and actively uses experimentation and iteration to solve problems.
- Empathizes with others and keeps the customer/user at the center of every decision.
- Sees the bigger picture—and finds connections between seemingly unrelated points and processes.
- Feels comfortable with ambiguity and uncertainty.
- Can express themselves clearly (whether using visual thinking or not) and engages in open, meaningful dialogue.
- Co-creates and collaborates with individuals and teams.
- Has a growth mind-set, which is a passion for learning rather than a hunger for approval.[5]

Empathic • Holistic • Open-minded • Courageous • Visual • Collaborative

2

DISCOVER YOUR NEW GROWTH OPPORTUNITY

THE FIRST PRINCIPLE OF FINDING NEW GROWTH IS THAT YOU'RE ALWAYS BETTER OFF GOING AFTER CUSTOMERS WHO ARE UNDERSERVED OR NEGLECTED.

—**DAVID BELL** *Professor*

Where to look for new growth opportunities? There is a growing body of evidence[1] suggesting that companies focusing on customers as the starting point of their strategic innovation efforts develop more successful and innovative offerings than those companies that start by searching inside their own organization. Why is that? Only by gaining a deep understanding of customers, their true needs and expectations, as well as their satisfaction or dissatisfaction with current offerings, will you gain the insights needed to develop solutions that customers really will want to buy. Most companies, unfortunately, don't know why customers do or don't do business with them in the first place.

In addition to understanding your customers, it still pays to have a sound, solid understanding of what your goals are, where your company is coming from, and which strengths it possesses that could be leveraged for new growth. And so, in order to discover opportunities for new growth and strategic innovation, focus on the following steps: (1) Understand customers and noncustomers; (2) Understand your company; and (3) Frame your growth initiative.

UNDERSTAND YOUR CUSTOMER AND NONCUSTOMER

THE CONVENTIONAL APPROACH to identify opportunities dictates that you analyze the current situation, define your objectives, and develop an approach on how to bridge the gap. We suggest looking outside your company, focusing on better understanding your customer and noncustomer, before looking at your company and framing your growth initiative.

Why? Simply because your search for opportunities should not be limited by your current position, assumptions, and any objectives you might have, before exploring the universe of opportunities that waits to be discovered.

Your journey to find your firm's new growth opportunity begins with the customer, but not by using traditional customer segmentation or market research methodologies. Growth arises from a deep understanding of your customers and noncustomers, their needs and expectations, what drives their choices, and the barriers to consumption and hurdles to satisfaction that hinder these needs from being fulfilled in a satisfactory manner.

When looking at customers, most companies focus on their best and often most satisfied customers, asking them what the company could do to make them even happier. Lufthansa, for example, interviews its first-class passengers and the members of the top tier of its frequent-flier program every quarter, asking them how satisfied they are with its services and what the airline could do even better. While we applaud these activities to keep one's most valued customers coming back, focusing on them is unlikely to result in exponential growth. They are already your best customers, so clearly they think you've got it right. Increasing your share of their wallet will take you only so far. Instead, we suggest taking a different track, by looking to a set of four groups of promising possible customers and noncustomers.[2]

MOST COMPANIES, UNFORTUNATELY, DON'T KNOW WHY CUSTOMERS DO OR DON'T DO BUSINESS WITH THEM.

CUSTOMER AND NONCUSTOMER

THE UNSATISFIEDS

The first group of customers to explore for new growth we call **unsatisfieds**, those people who are not contented with your offering. These customers may not buy your product at all, or they might consume it in small quantities, but only because there is no alternative (if there were an alternative, they would be the first to switch).

Imagine if a highly satisfying product, service, or customer experience could be offered to these customers.

Two companies that have turned unsatisfied customers into advocates are Nordstom and Zappos. Their strategy is based on delivering exceptional service and exceeding expectations. For example, Nordstrom's exceptional customer service comes primarily as a result of its attention to detail as it applies to the customer experience and the level to which it empowers its employees. Employees are given this guideline for their behavior: "Nordstrom Rules: Rule #1: Use best judgment in all situations. There will be no additional rules." Zappos also has legendary status when it comes to its customer service. The company's total devotion to providing quality service and listening to its consumers has paid off. It has a legion of loyal customers who evangelize about the outstanding service they receive from Zappos, which enjoys having 75 percent of purchases coming from repeat customers.

These firms have proven that going out of your way to accommodate customers' needs makes them feel important, respected, and in control. Converting an unhappy customer into a delighted one yields an incredibly loyal customer and brand evangelist.

CUSTOMER TYPES

REFUSERS
Aware but don't see the relevance.

UNSATISFIEDS
Aware but don't like the offering.

UNEXPLORED
Completely unconsidered audience.

WANNABES
Appealing but out of reach.

CUSTOMER AND NONCUSTOMER

THE WANNABES

The second group of prospective customers is **wannabes**, who are interested in your offering but cannot consume it for various reasons. Wannabes have barriers to consumption to overcome. Often, it is simply personal circumstances such being unable to afford your product or to gain access to your services. Or perhaps these customers may not possess the necessary skills to buy or consume your product, or to engage in any business activities with your company. An example of a company that overcame this challenge is the Belgian theater chain, Kinepolis, which realized that there was a large market of parents who would love to go to the movies, but the hassle of getting their children cared for was a barrier to ticket purchases. So, the company designed a program called Kinepolis Kids, providing activities and entertainment for children while their parents enjoyed a movie.

THE REFUSERS

The third group of potential customers we call **refusers**. These folks know about your product or service but simply refuse to buy and consume it. Reasons for refusal can be seeing your offering as irrelevant or without value to them, or as being too expensive; or perhaps they perceive it as too complicated to use or environmentally unfriendly. Younger generations often refuse to buy cars, for example, as they no longer perceive the car as a status symbol and they do not feel it is necessary to own a car; they refuse to invest money in a car they seldom need to drive. Businesses like Mercedes-Benz's Car2Go or Zipcar are built on addressing these refusers and designing an offering that nevertheless appeals to them and their occasional need for car transportation.

THE UNEXPLORED

The last group of potential customers is the **unexplored**. These are prospects that your industry has never thought of serving. The reasons you have never thought of serving these customers might be that your firm always thought of them as uninterested in your offering, belonging to a different industry, or not being profitable enough for your company. Kim and Mauborgne[3] cite the example of tooth-whitening products, which for years had been thought of as the strict domain of dentists. Then oral-care consumer product companies looked at the needs of some noncustomers and found a latent demand waiting to be fully mined by delivering safe, high-quality, low-cost tooth-whitening solutions. Another example is the European media company, ProSiebenSat.1. Start-ups and small and medium-sized companies had always been considered unprofitable and unable to afford TV advertising; hence nobody ever considered them as a potential customer segment.

The boundaries of these categories are not hard and fast. From a customer's perspective, he or she might be a wannabe, but from your company's perspective, that customer might be an unexplored, because you always considered the customer group to be unprofitable for your business. It does not matter. The important thing is that you think about these various customer and noncustomer categories and start exploring the dormant opportunities that lie within them.

How can you tap the potential that lies within these customer groups?

UNDERSTAND CUSTOMER NEEDS, EXPECTATIONS, AND CHOICE

Developing new offerings that will enhance your growth potential requires you to clearly identify the new prospective groups of customers, as well as understand what drives their choices, what their needs are, and what they expect of a good solution to fulfill those needs.

The job-to-be-done framework is helpful in understanding customer needs. The idea of focusing on the job to be done was originally developed by the late Harvard Business School professor Theodore Levitt, who is credited with saying, "People don't want a quarter-inch drill; they want a quarter-inch hole." The concept was later picked up and further developed by consultant Tony Ulwick, and popularized by Harvard Business School professor Clayton Christensen.[4]

The basic idea is that customers do not buy products because they want to own the product, but because they have an objective they would like to fulfill with that product. The product is a means to an end. Expanding on Levitt's thinking, people don't especially want a hole, but they do want to hang a picture, for example. Understanding this need led German chemicals company, Tesa, to develop self-adhesive hooks and strips to replace nails and screws, saving walls from holes and the customer from having to buy an expensive drill, the hassle of drilling into the wall, having to clean up the dust, the pressure of getting the holes aligned, and so on. So, by job to be done, we mean the problem of a given situation that needs to be resolved to a desired outcome. Once you understand the need and its dimensions and characteristics, along with the process to accomplish it, you can identify and design a growth opportunity. The essence of your growth strategy will be to satisfy your customers' needs and desires in a way that maximizes value for them, your organization, and other relevant partners.

So, should you just ask existing customers what they want and need? The automotive industrialist, Henry Ford, is credited with saying, "If I had asked people what they wanted, they would have said faster horses."

Instead of simply asking customers what they want, we suggest inquiring about what they want to achieve, what their needs are, which solution they use and why, and what they would expect from a good or better solution. Understanding customer needs is attained by focusing on the objective to be achieved, the outcome to be attained, the customer experience, and the process the customer goes through in order to come to this outcome. Besides investigating the underlying need, also look at which solutions, products, and services customers currently turn to in order to satisfy these needs, and examine their motivation for using these instead of others. What drives their choices?

Henry Ford's customers might have wanted faster horses, but they probably also wanted something that was a bit more comfortable and convenient, needed less maintenance, and possibly was cheaper. Observing customers, trying the products and services for yourself, and living the customer experience often reveal characteristics like these, which open up new opportunities.

For example, Clayton Christensen[5] tells about a fast-food company that learned that the reason early morning customers

DISCOVERING THE NEED FROM THE CUSTOMER'S PERSPECTIVE

These questions will help you uncover the essential need of the customer:

- What is the need, objective, or job to be done?
- What is the outcome the customer wants to achieve?
- What are the characteristics of the outcome?
- What is important for the customer?
- Where, when, and under which circumstances does the need emerge?
- Why does the need emerge?
- Why is the need important to the customer?

DISCOVERING DRIVERS OF CHOICE

These questions will help you to understand the factors and drivers influencing the customer's choice.

- Where, when, and under which circumstances does the customer buy, use, and consume your offering? Where, when, and under which circumstances is it bypassed?
- Which products and services does the customer consume when the need arises?
- Why does the customer buy these products or services?
- Which alternative offerings do the customers and noncustomers consider, buy, and use? Why these? Under which circumstances do they turn to these alternatives?
- Which products and services do the customers not use? Why not?
- How do customers use the various offerings available?

bought milkshakes had nothing to do with the flavor, size, or cup/straw design. Wanting to increase the sales at its stores, the company had someone observe customers. The observer saw that in the morning most customers would leave the store with a milkshake in hand. Asked about why they bought a milkshake to go (motivation) and what they did next (circumstances), the customers revealed that their need was to have a drink and activity that would make their long morning commute more interesting and would satisfy their stomach until noon. And the milkshake did a better job at fulfilling these needs than doughnuts, sweets, fruits, or other alternatives.

You might want to look not only at the need, but also at the circumstances when a product is being purchased or used. In the milkshake example, the circumstances were the early morning, and people leaving the store with their milkshakes in hand and getting into their cars.

Understanding why and when customers buy a certain product instead of another one opens up new ways of segmenting your market. Instead of looking at typical segmentation criteria like age, gender, income, and so on, the milkshake customers might be segmented into "morning commuters" and possibly something like "afternoon snack buyers in need of an energy boost." Segmenting according to needs and circumstances not only renders your segments a lot less complex, but it also enables you to create very targeted value propositions. Going to a restaurant might mostly serve the need to get food, but it often is also about socializing, spending a nice evening with friends, or having a romantic time with your spouse. A key motivation for using a smartphone is mostly to obtain or convey information, but sometimes it is just to kill time while sitting in airports, waiting in line, and so on to avoid being bored. Needs can hence be categorized as being functional, social, or emotional. For example, a car fulfills the functional need of transportation, but also the emotional need to feel safe or comfortable while doing so, and it might fulfill the social need for recognition by friends and peers.

A great way to visualize the set of choices, needs, and experiences of fulfilling them is through customer journey mapping. Customer journey mapping helps structure what customers actually want to achieve and the various steps they have to take to reach their objectives. Typical steps in the customer journey include realizing there is a need, becoming aware of potential solutions, evaluating and selecting a specific product or service, buying it, taking it home or getting it delivered, using it, maintaining it, and possibly also disposing of it.

Each step in the customer journey bears the potential to enhance the customer experience, creating value for the customer; to get the job done; and, if you do it better than anybody else, to increase your potential for growth. Once you understand your customers' needs and the experiences they have when trying to fulfill those needs, you can investigate what stands in their way to having a satisfying customer experience.

THREE TYPES OF NEEDS

FUNCTIONAL
Drill a hole.

EMOTIONAL
Hang a picture of loved ones.

THREE TYPES OF CUSTOMER AND NONCUSTOMER NEEDS

SOCIAL
Bring family together.

NEEDS BASED CUSTOMER JOURNEYS

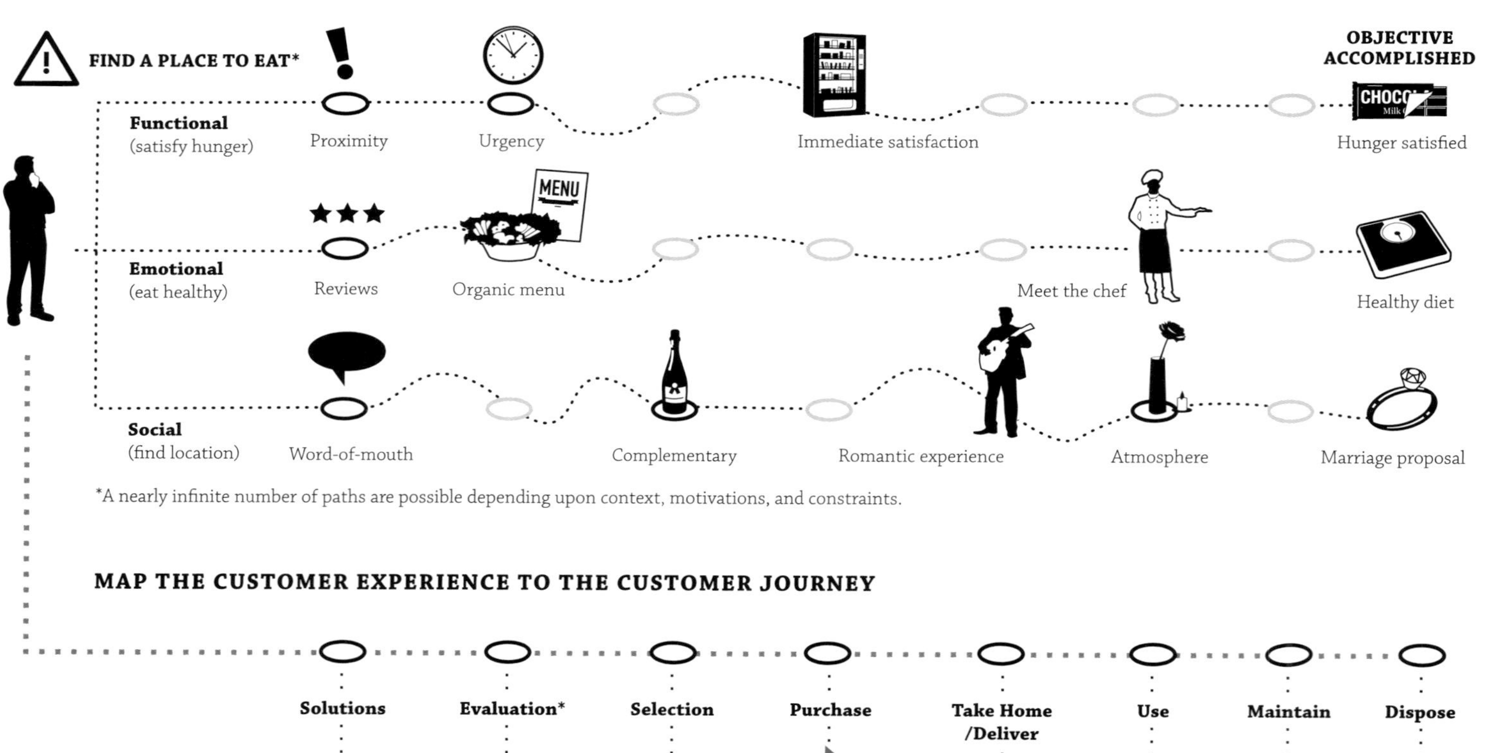

*A nearly infinite number of paths are possible depending upon context, motivations, and constraints.

MAP THE CUSTOMER EXPERIENCE TO THE CUSTOMER JOURNEY

Solutions | Evaluation* | Selection | Purchase | Take Home /Deliver | Use | Maintain | Dispose

Positive (+)

Negative (–)

BARRIERS TO CONSUMPTION — HURDLES TO SATISFACTION

*During evaluation, expectations for the post-purchase experience are established.

IDENTIFY BARRIERS TO CONSUMPTION AND HURDLES TO SATISFACTION

Having gained a deep understanding of your customers and noncustomers and the needs they have, you must figure out what is keeping the customer segments you identified from accomplishing the job to be done by utilizing the product or service, and what is keeping them from a positive customer experience.

The customer experience journey starts from the moment customers encounter their need and become aware of your product or service, and continues all through the buying cycle until the product or service is no longer used. All along this path, customers will encounter barriers, challenges, and roadblocks. To convince existing customers and gain new ones, you need to remove those obstacles.

Be sure to distinguish between **barriers to consumption**, which keep customers from buying and/or using your offering in the first place, and **hurdles to satisfaction**, which occur once your customer is buying and/or using your offering.

Typical barriers to consumption include:

- **Financial barriers:** Simply not being able to afford your offering.
- **Time barriers:** Not having enough time to go through the whole customer journey; for example, it might be too time-consuming to search for your offering or get it delivered, or delivery times might be too long.
- **Resource barriers:** Not having the resources necessary to find, evaluate, buy, transport, store, and maintain your offering; for example, customers might not have a car large enough to transport the new TV set home, or business customers don't have the necessary warehousing capacity to store the minimum amount of items you require them to buy.

CUSTOMERS AND NONCUSTOMERS IDENTIFY BARRIERS TO CONSUMPTION AND HURDLES TO SATISFACTION

- **Skill barriers:** Buyers might not have the skills needed to operate your product, or your online ordering system might be too complicated for them to get through.
- **Access:** Buyers may not be able to gain physical access to your offering.
- **Risk:** Your products or services may be outside your buyers' risk profiles; for example, relatively high cost or investment for the user increases the risk of buying online without having seen the product, booking a hotel where one hasn't stayed before, and so on.
- **Knowledge:** Buyers simply may not understand the product or have the knowledge or education to appreciate your product or service, or they lack the knowledge to make an informed choice.

Barriers to consumption can easily be uncovered by asking noncustomers why they are not choosing your offering, or what other companies they are buying from and why.

How can you identify noncustomers? If your customers are businesses, you probably have an idea which companies are not buying from you that you would nevertheless like to have as customers. Often you know that customers buy certain products from you, while not buying others.

You can also look at other industries or strategic groups within your industry to learn about alternative offerings that customers turn to in order to fulfill their needs. If you are in the hospitality business, for example, you might want to query customers from other hotels in your category, but also make sure to ask why customers trade up or down; why they turn to offerings like couch surfing, youth hostels, or offerings like Airbnb; or why they simply stay at home. (By the way, typical noncustomers for a hotel are people living in the same city.) Look at your competitors and ask their customers why they are not buying from you.

Hurdles to satisfaction typically arise around a less than satisfactory customer experience (e.g., having to wait in line at airport security, not sitting comfortably on the plane, etc.). In essence, every touch point between the customer and your company and your offering provides an opportunity for increasing either the customer's satisfaction or the cause of disgruntlement.

In general, hurdles to satisfaction can be expressed in terms of the three types of needs: functional, social, and emotional. Think of how well the offering fulfills each of the needs in these categories and what stands in the way of meeting these needs. Typical hurdles to satisfaction for these categories include: the offering does not fulfill the need, it is too complicated to use and creates frustration, it presents a risk for the customer, it evokes a bad conscience, it takes a lot of time to learn to use and operate, it is boring to use, and so on.

Hurdles to satisfaction can also be viewed through the customer journey lens. At each touch point, ask how difficult, costly (both financially and emotionally), and risky, or how easy, convenient, and affordable it is for the customer to move along the customer journey.

Typical hurdles to satisfaction include:

- **Awareness and selection:** Offering is difficult to find and compare; it is complex; decision making is difficult, or too many variants make it difficult to decide which one to buy.
- **Purchase:** Difficult to find a point of purchase; difficult ordering process; inconvenient payment options, or delivery hard to organize.
- **Usage:** Complicated usage requiring special expertise or skills, or demanding unreasonable effort or compromises from the user.
- **Supplements and maintenance:** Offering requires supplements for it to be operated reasonably well; supplements are expensive and need to be bought separately; maintenance is costly and can be done only by trained personnel.
- **Disposal:** Using the product creates a lot of waste, which is hard to dispose of; the product cannot be easily disposed of; disposal creates cost (financial, time, emotional) and requires special treatment; users don't know how or where to dispose of the product after it has been used.

To identify hurdles to satisfaction, it is again important to have understood the customer's needs. Consider visiting a restaurant. If your need is primarily functional (i.e., satisfying your hunger), long waiting times are a hurdle to satisfaction. If, on the other hand, your need is social (i.e., you want to spend time with friends), quick service can be disturbing, as you won't have time to chat with your pals. It is only through understanding what customers want to achieve and why they are consuming your offering (or not) that you can identify the barriers to consumption and the hurdles to satisfaction and then design an exciting offering that will please them.

The two Inspiration case examples on the following pages illustrate how understanding your customers' needs, expectations, and frustrations with the current experience can lead to new growth strategies.

INSPIRATION

CARDINAL HEALTH ENHANCING THE CUSTOMER EXPERIENCE

Cardinal Health has a long history of finding and capturing opportunities through strong user-focused attention and unswerving focus on improving on its customers' experiences. The company started in 1971 as a food distributor in Columbus, Ohio. Four years after adding pharmaceutical distribution in 1979, it went public. Cardinal continued to expand its offerings and sold its food distribution arm in 1987. Cardinal Health is now ranked 26th in the Fortune 500, and is a $103 billion health care services company with 34,000 employees on four continents.

Cardinal Health exploited barriers and hurdles in the customer experience cycle of the hospital surgical kits delivery business. The average surgical tool kit varies greatly with the procedure and the doctor's preferences and requires some 200 products. Traditionally, thousands of different items must be stored in hospital stockrooms, handpicked before a procedure, and transported on a tray to the operating room. The process is expensive, time-consuming, and error-prone.

Cardinal developed an online ordering tool enabling surgeons to walk through their procedures in advance, picking the equipment and supplies they prefer. The multitude of items needed for a particular operation are then shipped to the hospital on the morning of the procedure in a sterile kit, organized in the precise sequence in which they will be used.

Cardinal enhanced the customer experience by offering a solution—in this case a digital portal that can be customized to customers' specific requirements—to make the ordering process more convenient and less complex, and, hence, to better meet its customers' needs. The growth came from careful observation of its customers' experience journeys, observing customer frustrations, and finding ways to help make their job to be done easier, while having a better experience.

KLINIK HIRSLANDEN DISCOVERING CUSTOMER NEEDS

Through careful listening to its ecosystem, Klinik Hirslanden, a Swiss private hospital, was able to evolve its business model to better address the needs of its stakeholders. When Dr. Daniel Liedtke became the new managing director of the hospital, he did two things. First, he conducted about 70 semistructured interviews with internal and external stakeholders of the hospital during his first 100 days in the office. Based on this assessment of the organization, he decided it was necessary to rethink its business model. To do so, he put his management team on a journey to understand who Hirslanden's customers were and why they chose Hirslanden instead of any of the roughly 300 other hospitals in Switzerland.

Using a mix of personal interviews with patients and doctors, patient feedback surveys, and secondary research available on performance and satisfaction drivers in hospitals, Hirslanden revealed three important insights. First, mostly the family doctors and other referring doctors chose Hirslanden, not the patients themselves. These referrers were defined as a new customer segment, and special care was taken to inform them about the services of the clinic and engage them more. Second, referrers chose Hirslanden because of its excellent services and reputation in areas of highly specialized medicine. As a result, a business model was developed that ensured more specialist surgeons would come work at Hirslanden. Third, Hirslanden learned that once the patient was in the hospital, nonmedical services like the quality of the food and personal services were much more important to the patient's satisfaction than the medical treatment itself. As a result, hospitality services were added to Hirslanden's offering. (More details on Klinik Hirslanden's strategy will be shared in Chapter 3.)

HOW TO UNDERSTAND YOUR CUSTOMER AND NONCUSTOMER

GATHER INFORMATION

How do you go about exploring and understanding your customer and noncustomer? First, you will need to engage and gather information, make observations and create themes, and then transform findings into insights. Don't rely on traditional market research methodologies. Go out and talk to real customers, observe them, be a customer yourself, and live the experience. Personal engagement and interaction will be key to really discovering the needs, motivations, experiences, and frustrations of customers. We also suggest that you conduct interviews yourself, instead of hiring a research company. You can still supplement your own experiences with secondary research or quantitative data collected through larger-scale surveys. Customers will feel more valued if people from your firm get in touch with them.

Many activities can help you to gather information, beginning with interviews with individuals, groups, or experts. For some direct real-world observation of your customers you can go outside and watch customers in their environment. You can also tap into networks of stakeholders in your customer's community. Sometimes this can be an actual innovation network of people, institutions, or companies that exist outside of your firm. But, you might also have these networks inside your firm. Another more impactful observation technique involves immersing yourself in the environment in question, a.k.a., being a "fly on the wall," or experiencing the user journey firsthand by participating in the customer's actual process

The key point is to get outside! Seek inspiration in new places. Look at the set of experiences, needs, and behaviors associated with your opportunity. Consider where else and when

INFORMATION GATHERING ACTIVITIES

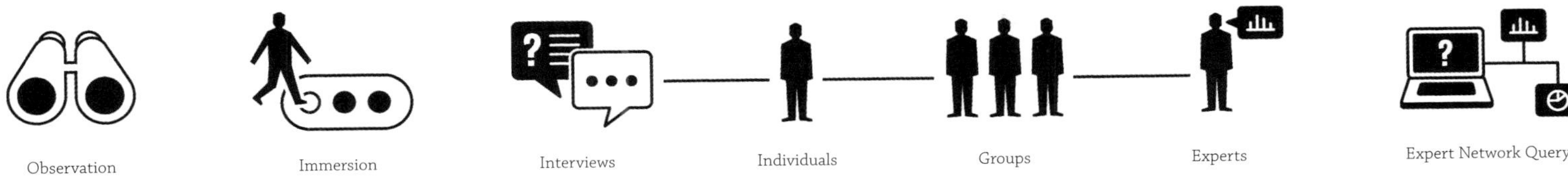

these sets of conditions occur, and go observe that situation. You might gain a completely new perspective and an invaluable insight. This active engagement is a critical aspect of the process of finding your opportunity.

MAKE OBSERVATIONS AND CREATE THEMES

Once you have explored your wide opportunity universe, create observations and synthesize them into key learning and themes through the Reveal Observation Headlines activity session. Observations are the key ideas that stood out from your fieldwork of gathering of information: simple stories, quotes from an interview, notes from field research. Curate your many ideas with your team, and organize a summary of points that can be used in your workshop. This part of the process can take a bit of time.

You will collect and share the stories from your fieldwork, and start making sense of all that information and inspiration. Then aggregate the most interesting observations of your field research into theme-based categories.

TRANSFORM FINDINGS INTO INSIGHTS

Once you, and your teammates, have collected all the information, you need to make sense of it and try to understand it. There is a host of activities you can entertain to make sense of the wealth of information you have gathered. To gain clear understanding of and empathy for the customer and noncustomer segments, you can develop persona maps for the customer types you observed. This is particularly useful in gaining insight and building your strategy. By leveraging personas, you can use storytelling and visualization to communicate the heart of the customer's needs, represent the barriers to be overcome and the hurdles to satisfaction, and articulate the core job to be done.

INSPIRATION

RIPPLE EFFECT GROUP A HUMAN-CENTERED DESIGN APPROACH

A provider of post-graduate medical education, which we will call, "The College," is responsible for the training and assessment of doctors to further their careers in a broad range of specialty fields. The College recognized that their educational offerings required a strategic review to enable the development of new and innovative approaches to learning with technologies, eLearning, in an increasingly complex environment.

The Ripple Effect Group was engaged to review and develop a new eLearning strategy for the future delivery and support of post-graduate programs, based on bringing value to its student doctors. The review covered three phases: discovery, sensemaking of the data and observations, and recommendations.

Phase 1: Discovery. In order to develop a deep understanding of the people involved and the complexity of medical education settings, a human-centered design research approach informed the critical elements in the first phase of the project.

Ripple Effect Group gathered insights through a rapid ethnographic study over a 3-month period. The research team was immersed with trainees and supervisors in their workplace environment, predominantly in hospitals. Techniques applied were direct observation of ward rounds, shadowing people throughout their day, informal discussions during meal breaks, attendance of lectures or other education sessions, review of materials, and participation in briefing sessions, which provided a rich set of raw data and experiences.

Phase 2: Sensemaking of the data and observations. The research team collated and debriefed the raw data using visual thinking techniques that mapped the experiences. The objective was to determine the specific nuances, constraints, and challenges in a hospital training environment, identifying opportunities where innovative techniques and technologies could provide a practical alternative to current methods. These visual assets were used during consultation review sessions to engage with the broader physician community and validate research findings against the various contexts and experiences.

- **Personas:** A series of authentic personas were constructed to represent characteristics of groups at different stages in their careers, outlining students' needs, barriers to adoption, and hurdles to satisfaction
- **Learning Journeys:** Scenarios relating to specific educational situations were created to illustrate the understanding of contexts and to guide the development of recommendations for the state's future learning programs.

Phase 3: Recommendations. The information gathered during Phases 1 and 2 were presented as detailed reports and further supported by analysis of medical education globally.

- **Current State Report:** detailed the research findings and analysis of the current state.
- **Best-in-Class Medical Education:** highlighted global best in class practices and initiatives.
- **Organizational Analysis Report:** detailed the theoretical application of the learning initiatives examined and made recommendations that identified current and future roles, resources, and assets required within The College to contribute to future changes.
- **Future of Learning Report:** a strategy that identified the core elements and actions to enable The College to transform their offerings. The recommendations and implementation steps presented to the executive sponsors was enhanced by the power of the visual assets to illustrate how the new initiatives would be experienced and delivered in a future scenario.

THREE KEY LEARNING POINTS

1. Human-centered design provided insights into people that could not be uncovered by traditional surveys, focus groups, or interviews.
2. Visual thinking and storytelling assets created from the research findings enabled people to understand and connect concepts that may have been too abstract to grasp in traditional written documentation.
3. Engagement across all levels of The College ecosystem ensured research findings were validated from more than one perspective.

OTHER OUTCOMES

The visual assets developed became an integral communication and engagement tool across The College. Working groups, committees, and educational designers are constantly reviewing new initiatives against the set of authentic persona to validate that all relevant aspects have been considered and addressed.

PERSONA MAPPING

Persona mapping builds empathy, gainings alignment with your team around the customer/noncustomer needs, goals, pain points, and motivations.

OVERVIEW

Persona mapping creates a picture of a customer or noncustomer to build empathy and better understanding of their needs and barriers.

TIME NEEDED

45–60 minutes

MATERIALS

Markers, sticky notes, and whiteboard

STEPS

1. Start by drawing the profile head of the target customer or noncustomer and give the persona a name and some information to identify them (i.e., job title, age). The profile can be as simple as a circle with an eye, nose, mouth, and ear. Alternatively, you can also post a picture of a typical customer.
2. Divide the working space around the "head" into sections that represent aspects of that person's sensory experience. Label them: thinking, feeling, seeing, saying, doing, hearing.
3. Give each participant a pad of sticky notes and marker. Instruct them to project themselves into the target customer. Keeping in mind the customer's experience, motivations, barriers to consumption, and hurdles to satisfaction as you delve into the context you want to explore (e.g., use of your product or purchase experience).
4. Have each participant write what the persona is seeing, saying, hearing or doing in relation to the context. Write only one idea/concept per note and place them in the appropriate section of the persona map. Work from the outside world (seeing, saying, hearing, doing). Next, move to thinking and feeling. This is intentional and the sequence is important. Start with the concrete—what is observable and tangible about their experience—and then move to the intangible.
5. Once persona discovery is complete, the entire group should review and discuss what they've learned. Look for clusters of similar information and try to align key points. Take photos of all the maps created.

Additional activity resources and templates can be found at www.theartofopportunity.net.

SOURCE: *Gamestorming:* Gray, Macanufo, and Brown

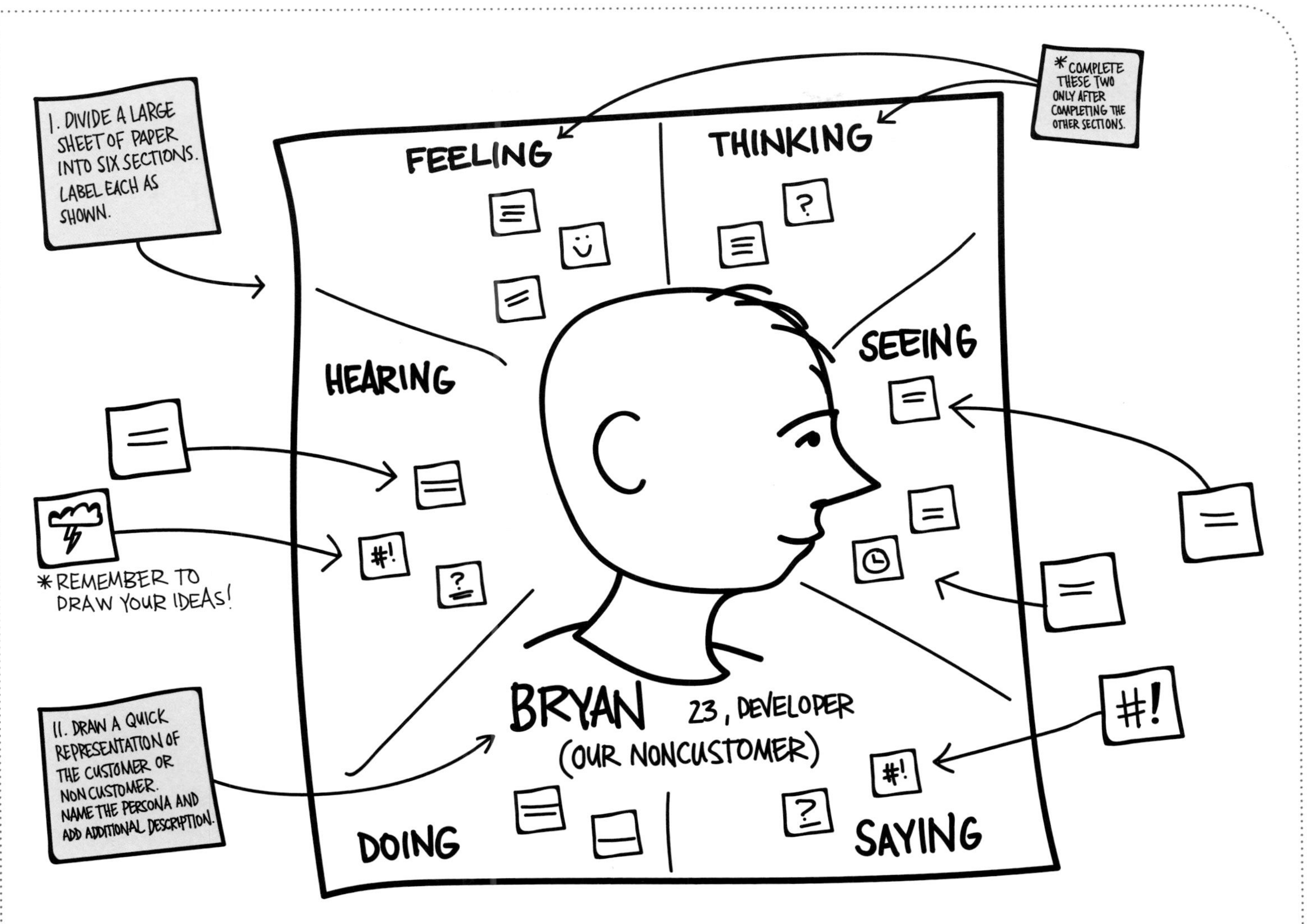

I. DIVIDE A LARGE SHEET OF PAPER INTO SIX SECTIONS. LABEL EACH AS SHOWN.
* COMPLETE THESE TWO ONLY AFTER COMPLETING THE OTHER SECTIONS.
FEELING
THINKING
HEARING
SEEING
* REMEMBER TO DRAW YOUR IDEAS!
II. DRAW A QUICK REPRESENTATION OF THE CUSTOMER OR NONCUSTOMER. NAME THE PERSONA AND ADD ADDITIONAL DESCRIPTION.
BRYAN
23, DEVELOPER
(OUR NONCUSTOMER)
DOING
SAYING

CREATE AN INTERVIEW GUIDE

Design an interview guide to gather a set of similar information from individuals, groups, or experts.

OVERVIEW

There is probably no better way to gain insight and gather information than to speak directly with people. This goes to the heart of a user-centered approach to best understand needs, wants, and barriers to purchase and use.

TIME NEEDED

90–120 minutes

MATERIALS

None but your creativity

STEPS

1. Create a set of questions that go to the heart of the customer's needs. You are not seeking opinions on feasibility at this point. Do not ask for product/service features either. Your primary objective is to gain a deep understanding of your customers by finding patterns of behavior, values, and needs of the customer or noncustomer.
2. As a rule of thumb, start your interview guide with a few easy, identifying questions (name, job, company, etc.). Then ask some broader open-ended questions that are a bit more unexpected; perhaps ask about future aspirations, their needs, circumstances, when the needs arise, which alternative solutions they turn to, why these and not others.
3. One successful technique involves creating some comparative concepts and giving the subject two scenarios. For example, say to the interviewee, "In Case A, you can obtain 'x' number of 'z' very quickly. But in Case B, you can obtain 'y' number of 'z,' but have to wait to obtain them. Which would you prefer?" This is especially valuable in getting a sense of the perceived value of an opportunity.
4. Share the interview guide with other team members for feedback, and then take it on a "test drive." After trying it out on a few interviews, adjust as needed. Take photos of all whiteboards.

Additional activity resources and templates can be found at www.theartofopportunity.net.

SOURCE: *Gamestorming*, IDEO's Human-Centered Design Tool-kit, *101 Design Methods*

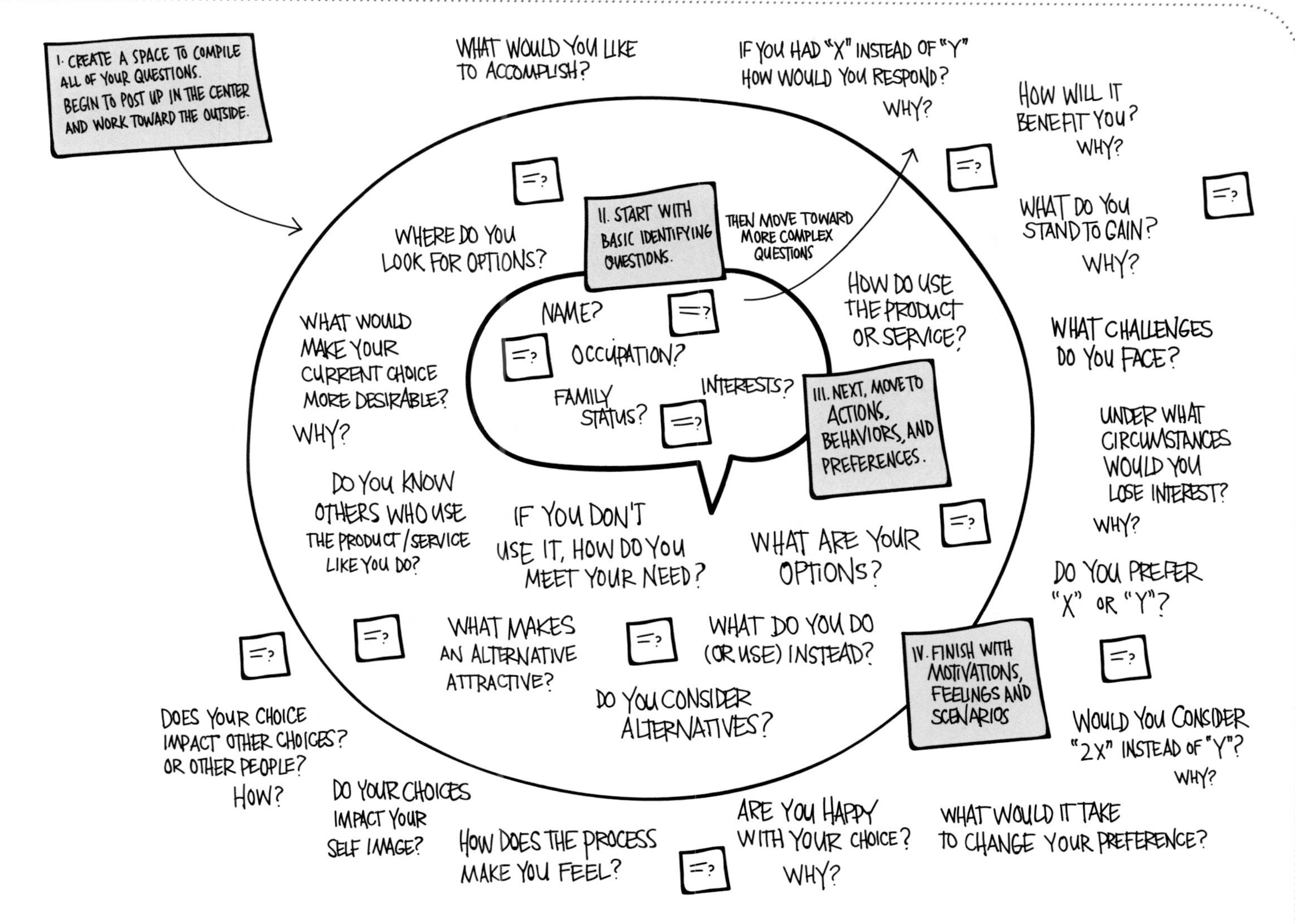
I. CREATE A SPACE TO COMPILE ALL OF YOUR QUESTIONS. BEGIN TO POST UP IN THE CENTER AND WORK TOWARD THE OUTSIDE.
II. START WITH BASIC IDENTIFYING QUESTIONS.
THEN MOVE TOWARD MORE COMPLEX QUESTIONS
III. NEXT, MOVE TO ACTIONS, BEHAVIORS, AND PREFERENCES.
IV. FINISH WITH MOTIVATIONS, FEELINGS AND SCENARIOS
NAME?
OCCUPATION?
FAMILY STATUS?
INTERESTS?
WHAT WOULD YOU LIKE TO ACCOMPLISH?
IF YOU HAD "X" INSTEAD OF "Y" HOW WOULD YOU RESPOND? WHY?
HOW WILL IT BENEFIT YOU? WHY?
WHAT DO YOU STAND TO GAIN? WHY?
WHERE DO YOU LOOK FOR OPTIONS?
HOW DO USE THE PRODUCT OR SERVICE?
WHAT WOULD MAKE YOUR CURRENT CHOICE MORE DESIRABLE? WHY?
WHAT CHALLENGES DO YOU FACE?
UNDER WHAT CIRCUMSTANCES WOULD YOU LOSE INTEREST? WHY?
DO YOU KNOW OTHERS WHO USE THE PRODUCT/SERVICE LIKE YOU DO?
IF YOU DON'T USE IT, HOW DO YOU MEET YOUR NEED?
WHAT ARE YOUR OPTIONS?
DO YOU PREFER "X" OR "Y"?
WHAT MAKES AN ALTERNATIVE ATTRACTIVE?
WHAT DO YOU DO (OR USE) INSTEAD?
DO YOU CONSIDER ALTERNATIVES?
WOULD YOU CONSIDER "2X" INSTEAD OF "Y"? WHY?
DOES YOUR CHOICE IMPACT OTHER CHOICES? OR OTHER PEOPLE? HOW?
DO YOUR CHOICES IMPACT YOUR SELF IMAGE?
HOW DOES THE PROCESS MAKE YOU FEEL?
ARE YOU HAPPY WITH YOUR CHOICE? WHY?
WHAT WOULD IT TAKE TO CHANGE YOUR PREFERENCE?

CUSTOMER JOURNEY MAPPING

A visualized journey map lets you see what is going on at each stage of a customer's, or noncustomer's experience.

OVERVIEW

Capture multiple levels of observation through a journey map. Aspects such as social, emotional, functional, and aspirational experiences can be seen over the time frame of the user's engagement.

TIME NEEDED

90–120 minutes

MATERIALS

Creativity, a whiteboard, markers, and sticky notes

STEPS

1. Give each participant sticky notes and markers. After identifying the customer or noncustomer to map, each participant identifies (and writes on sticky notes) all the activities throughout the customer experience, from initial awareness to the final action they take in their journey. What is the first step of awareness and attraction? What is the first touch point with the offering? Once engaged, what experiences occur? What happens at the stage when the customer leaves the experience? Is there a phase where the customer's experience is extended? Place the activities on a whiteboard. Group the activities into clusters (or stages) of related activities.
2. Name each cluster/stage and write it on a sticky note. Place the sticky notes as headers at the top of the whiteboard. Next, create rows to frame your journey with dimensions or lenses of perception. Examples of dimensions include: social, emotional, and functional; or pain points and moments of "wow." You can also include channel or place as a dimension. Write your dimensions on sticky notes to label each row.
3. For each row, participants will consider that dimension as it relates to each "stage" in the customer journey. What is happening in relation to that dimension? What are they thinking and feeling? Write ideas on sticky notes and post on each stage across the dimensions.
4. When complete, discuss as a team to review all the ideas. Remove duplicate sticky notes. Simplify and edit to create a visual map that's easy to see and understand.
5. Continue your team discussion to synthesize the key insights you have made while creating the customer journey. Craft your insights into succinct, memorable full-sentence statements. Take photos of all whiteboards.

Additional activity resources and templates can be found at www.theartofopportunity.net.

SOURCE: *The Art of Opportunity* authors and inspired by *Gamestorming, IDEO's Human-Centered Design Tool-kit, 101 Design Methods*

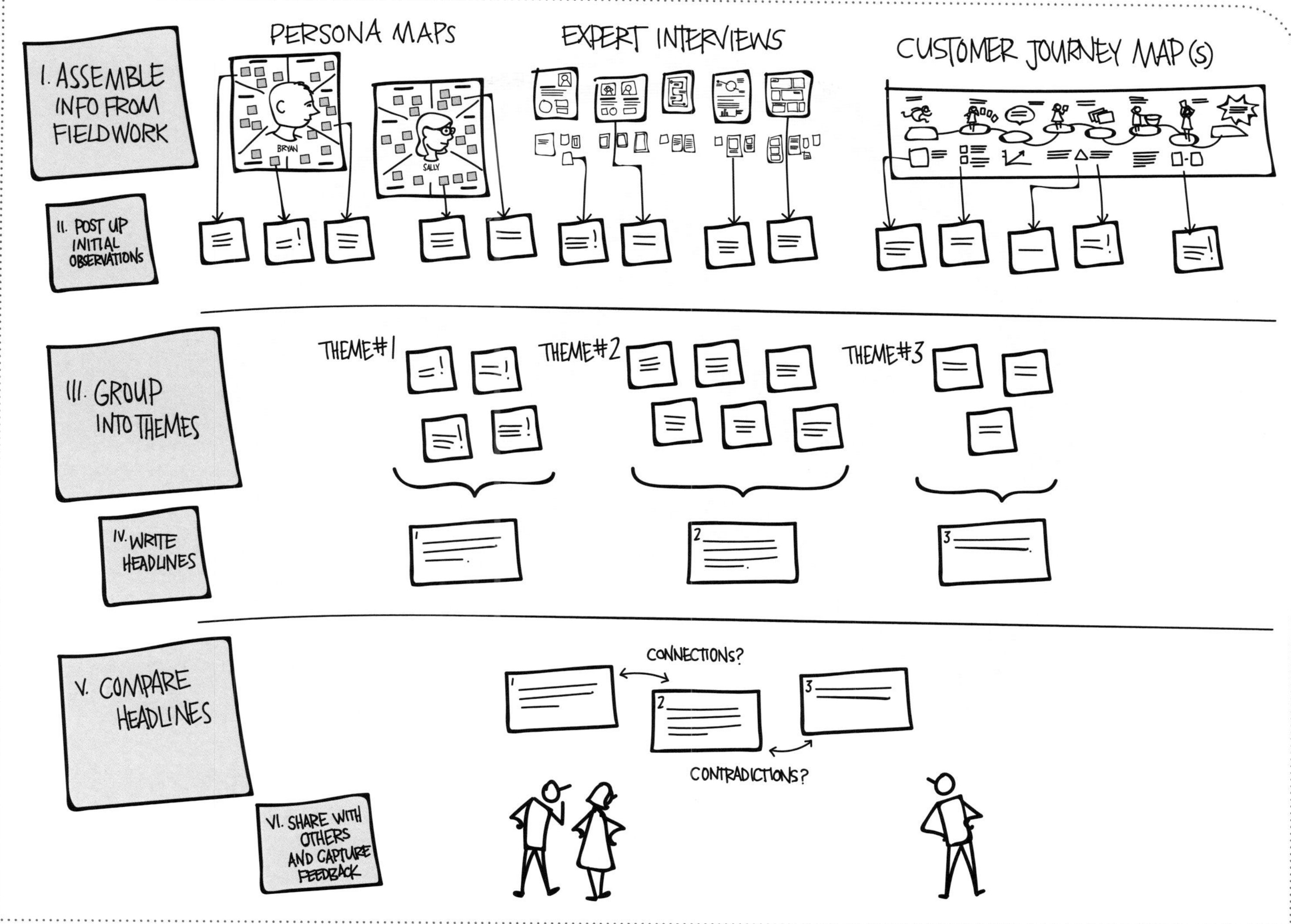
I. ASSEMBLE INFO FROM FIELDWORK
PERSONA MAPS
BRYAN
SALLY
EXPERT INTERVIEWS
CUSTOMER JOURNEY MAP(S)
II. POST UP INITIAL OBSERVATIONS
III. GROUP INTO THEMES
THEME#1
THEME#2
THEME#3
IV. WRITE HEADLINES
V. COMPARE HEADLINES
CONNECTIONS?
CONTRADICTIONS?
VI. SHARE WITH OTHERS AND CAPTURE FEEDBACK

REVEAL OBSERVATION HEADLINES

Make sense of all the information you gathered in your observation fieldwork to find themes and patterns.

OVERVIEW

Create general understanding and find themes and patterns from your fieldwork research, then build observation headlines so your team can design growth opportunities and ideas.

TIME NEEDED

2–3 days in preparation; 60 –120 minutes in workshop

MATERIALS

All the information, photos, and documents from your observation fieldwork; whiteboard, sticky notes, and markers

STEPS

1. Collect and share the stories from your fieldwork in a large open space. Preparation may take a few days. Have enough space to spread out and easily review the photos, stories, audio, notes, and anecdotes from your observation. You may organize by ways you gathered information (e.g., individual interviews, expert groups, and immersive observation).
2. Study your assembled observation fieldwork. Every participant should choose 3–4 points they find most interesting or surprising and write them on sticky notes. Share with the team one at a time. Are there any major "wows?" From the collection of observations, identify common themes and group those notes together.
3. For each theme create a headline. You can do this as individuals or in groups. Each headline should be a complete sentence or statement summarizing the theme. Share with the team. Select the most interesting headlines.
4. Study your headlines. Look for connections and links among them. Are there anomalies or contradictions? Are some interrelated? What surprised you or did you find most curious? What's missing? Is there a dominant theme to the headlines?
5. Assemble your themes and headline statements on a whiteboard and have another team from your firm review it. Are there any additional insights or statements that can be developed? Revisit all your observations and themes, synthesize and assemble the insights you have made, and craft them into succinct memorable full-sentence statements. Take photos of all whiteboards.

Additional activity resources and templates can be found at www.theartofopportunity.net.

SOURCE: *The Art of Opportunity* authors

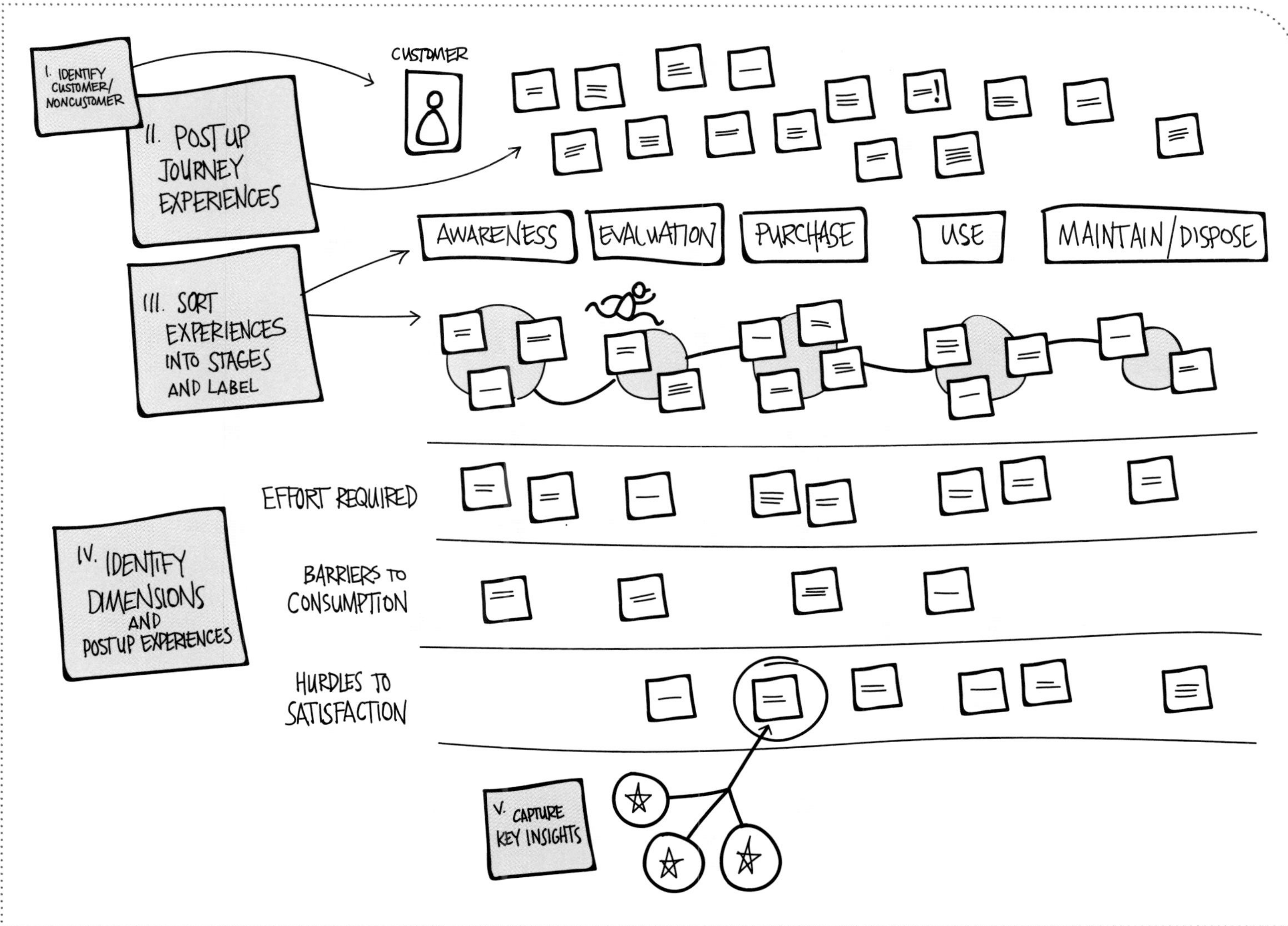
I. IDENTIFY CUSTOMER/ NONCUSTOMER
CUSTOMER
II. POSTUP JOURNEY EXPERIENCES
III. SORT EXPERIENCES INTO STAGES AND LABEL
AWARENESS
EVALUATION
PURCHASE
USE
MAINTAIN/DISPOSE
IV. IDENTIFY DIMENSIONS AND POSTUP EXPERIENCES
EFFORT REQUIRED
BARRIERS TO CONSUMPTION
HURDLES TO SATISFACTION
V. CAPTURE KEY INSIGHTS

UNDERSTAND YOUR FIRM

LOOKING AT YOUR CUSTOMERS and noncustomers can be thought of as an external search approach to uncovering growth opportunities, but looking at your company is more of an internal exploration tactic. While looking for opportunities through an internal lens, the key still is to discover opportunities for growth from existing and new customers. To do this, you will need to reflect on your current offerings and most valuable assets, what needs they fulfill that could be interesting for noncustomers, and how they could be leveraged to address different needs altogether.

Understanding your company and discovering its strengths are usually not big challenges. Once you have understood why customers come to you, you will already have a good sense of what you excel at doing. And this understanding will usually go beyond merely knowing which products and services customers like, to understanding the underlying capabilities and competencies that make your company special.

Most people, for example, would say that if they buy Apple products, it is because of their design and simplicity. That is the core of Apple. And its capabilities are primarily in these areas. Once you understand these underlying assets, you can leverage them into other business areas. Apple went from laptops and computers into music players, smartphones, and tablets applying these capabilities.

Amazon is all about convenience (easy ordering, home delivery), low price, and low risk (as you can ship almost anything back). In addition, Amazon possesses capabilities and assets that can be leveraged into other areas. The online bookshop, along with its convenience and low risk, for example, has easily been leveraged to sell other goods besides books. At the same time, its logistic capabilities and warehousing assets have been leveraged to offer fulfillment services to small and medium-sized businesses, meeting their needs for such services. In addition,

YOUR FIRM
Understand your firm's underlying power.

Amazon leverages its access and knowledge about its online customers, offering these businesses the opportunity to sell on Amazon's platform. The online business model requires a vast amount of servers and data processing power, which Amazon has leveraged to offer Amazon Web Services, and hence fulfill the needs of a new customer group completely unrelated to the online shoppers.

McDonald's leveraged its most valuable asset, real estate, to create a new offering, the McCafé, catering to a different need and different customer segment. Barnes & Noble and other retailers rent out space in their stores to Starbucks and other bar and restaurant brands.

The design consultancy, IDEO, leverages its design thinking methodology in areas such as education and solving social issues.

RESOURCES AND CAPABILITIES

THINK ABOUT THE STRENGTHS, capabilities, and resources of your business that you could put to use and leverage to create new businesses. What are the capabilities at the core of your products and services? What are you particularly good at doing? Which valuable assets do you possess? What customer needs could be satisfied with these? Which needs do your products, services, and assets fulfill? Which noncustomers might have such a need?

Typical capabilities and resources to think about:

- Unique skills in sales, manufacturing, research and development (R&D), logistics, recruiting, design, customer service, and so on.
- Unique customer access, distribution channels, etc.
- A loyal customer (fan) base.
- Knowledge, information, and data about customers, technologies, competitors, markets, and so forth.
- Access to and relationships with customers, suppliers, partners, and others.
- Effective and efficient processes.
- Brand, image, and reputation.
- The ecosystem your company is embedded in.

Typical assets and tangible resources to think about:

- Real estate, land, buildings, etc. in special valuable locations.
- Unique and special equipment, machinery, and so forth.
- Financial resources, revenues, low cost, easy access to cheap financing, and so on.
- Hardware.
- Human resources.

YOUR FIRM

Discover your strengths.

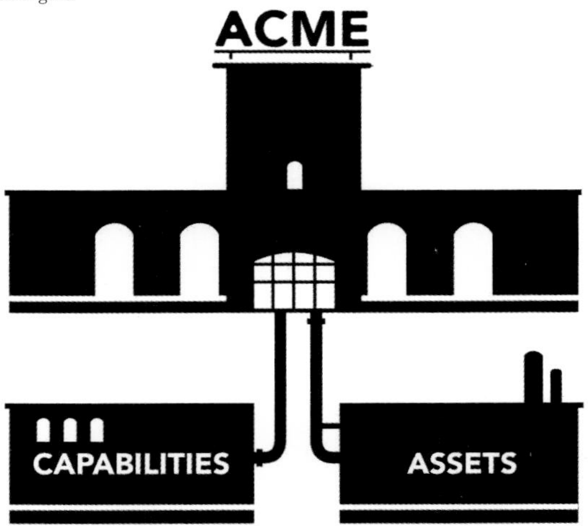

ACME

CAPABILITIES

ASSETS

CAPABILITIES

Find your capabilities.

RESOURCES

Leverage your assets and resources.

Looking at ecosystems can provide an interesting lens to discover new growth opportunities. Ecosystem opportunities lie mostly in facilitating the collaboration and exchange of value between ecosystem players. This can, for example, be achieved through creating new technology platforms like Uber or Airbnb that tie the ecosystem players together.

Other ecosystem opportunities lie within better understanding of your organization's existing ecosystem and creating strategies and, above all, business models that leverage the core capabilities of all ecosystem players while offering increased value to all of them, as we will discuss in more detail in Chapter 3 (hint: look for the Klinik Hirslanden Inspiration case study).

INSPIRATION

CAESARS PALACE FINDING NEW REVENUE FROM YOUR RESOURCES

Caesars Palace was built by Jay Sarno in 1966 and was, at the time, one of the most extravagant undertakings in Las Vegas. While it has changed owners over the years, the luxury hotel had built its reputation by catering to high rollers and presenting superstar entertainment and special events.

In the 1980s, the lavish hotel-casino was viewed as the "queen of the Las Vegas strip." The mission of its entertainment department was to attract customers to the shows and nightclubs, where they would, of course, pass the world's best concession operation—the casino. As such, the management's view was that it was okay to lose money on entertainers' fees, since those investments would be recouped on the cash take from gambling.

However, the hotel longed for revenue growth and looked to its directors of entertainment to come up with innovative ideas to grow its customer base and attract a wider audience. The management team examined the makeup of its business model, including its activities, assets, and resources, exploring what it did well, and came up with a unique solution.

Caesars had a huge parking lot and had already erected a stadium for boxing events. It had a stellar reputation for entertainment, and had operational knowledge and resources to handle security, food service, and major event production. Based on these advantages, the entertainment team executives decided to create Las Vegas' first outdoor concert series using the underused parking lot as a venue. Venue specifications were researched, talent was secured, and the hotel personnel developed an operational plan to launch the series.

The "Concerts Under the Stars" series was a huge success. Thousands of people attended, ate food, and gambled. Caesars had attracted a new customer segment and, for the first time, the entertainment department was not a loss leader, but became a profit center.

FOR THE FIRST TIME EVER, THE ENTERTAINMENT DEPARTMENT WAS NOT A LOSS LEADER FOR THE HOTEL, BUT BECAME A PROFIT CENTER.

RESOURCE SPARKS

THE FOUR DIMENSIONS FOR MAPPING AND ASSESSING RESOURCES

Following is a set of "Resource Sparks" to help you explore the resources of your firm using the four VRIO dimensions (valuable, rare, costly to imitate, and organized) to assess their unique competences and innovative capabilities.[6] You will find that optimizing your resources to be valuable, uncommon, and difficult to copy, and then being well-organized to capture the value of your resources, will deliver significant competitive advantage.

FIND YOUR VALUABLE RESOURCES:

- Which activities lower the cost of production without decreasing perceived customer value?
- Which activities increase product or service differentiation and perceived customer value?
- Has your company won an award or been recognized as the best in something (most innovative, best employer, highest customer retention, or best exporter)?
- Do you have access to scarce raw materials or hard-to-get-into distribution channels?
- Do you have a special relationship with your suppliers, such as a tightly integrated order and distribution system powered by unique software?
- Do you have employees with unique skills and capabilities?
- Do you have a brand reputation for quality, innovation, and customer service?
- Do you do perform any tasks better than your competitors do? (Benchmarking is useful here.)
- Does your company hold any other strength compared to rivals?

FIND YOUR RARE RESOURCES:

- How many other companies own a resource or can perform capably in the same way in your industry?
- Can a resource be easily bought in the market by rivals?
- Can competitors obtain the resource or capability in the near future?

FIND YOUR COSTLY-TO-IMITATE RESOURCES:

- Can other companies easily duplicate a resource?
- Can competitors easily develop a substitute resource?
- Do patents protect it?
- Is a resource or capability socially complex?
- Is it hard to identify the particular processes, tasks, or other factors that form the resource?

ORGANIZE TO EXPLOIT RESOURCES:

- Does your company have an effective strategic management process within the organization?
- Are there effective motivation and reward systems in place?
- Does your company's culture reward innovative ideas?
- Is an organizational structure designed to use a resource?
- Are there excellent management and control systems?

METHODS TO UNDERSTAND YOUR FIRM

WE'VE CREATED A SET OF ACTIVITIES you can apply to better understand your firm, its strengths, weaknesses, and possible opportunities. The first activity is mapping your tangible and intangible resources, which brings you to the essence of what your firm does well and the unique assets it has to leverage. A tangible resource is a three-dimensional physical item, something you can touch. An intangible resource cannot be detected by any of the human senses. Just like with cooking a meal, it is important to know what ingredients you will use to create your opportunity and how much time you have to work within.

The second activity in your process is mapping your firm ecosystem. James F. Moore first defined a business ecosystem in 1993 as "an economic community supported by a foundation of interacting organizations and individuals—the organisms of the business world."[7] Understanding your ecosystem can have a dramatic impact on how you organize, operate, and plan for the future. By mapping your ecosystem, you can help your firm improve its effectiveness, manage risk, and develop new innovative strategies for growth. As important, using your ecosystem model, you can also identify key places to explore and test your opportunity. While some will see this exercise as developing a set of restrictions, there is a growing body of work that views constraints as providing the inspiration, creativity, and insights for innovation and beautiful outcomes.

OPPOSITE: Walt Disney Studios Ecosytem, 1957

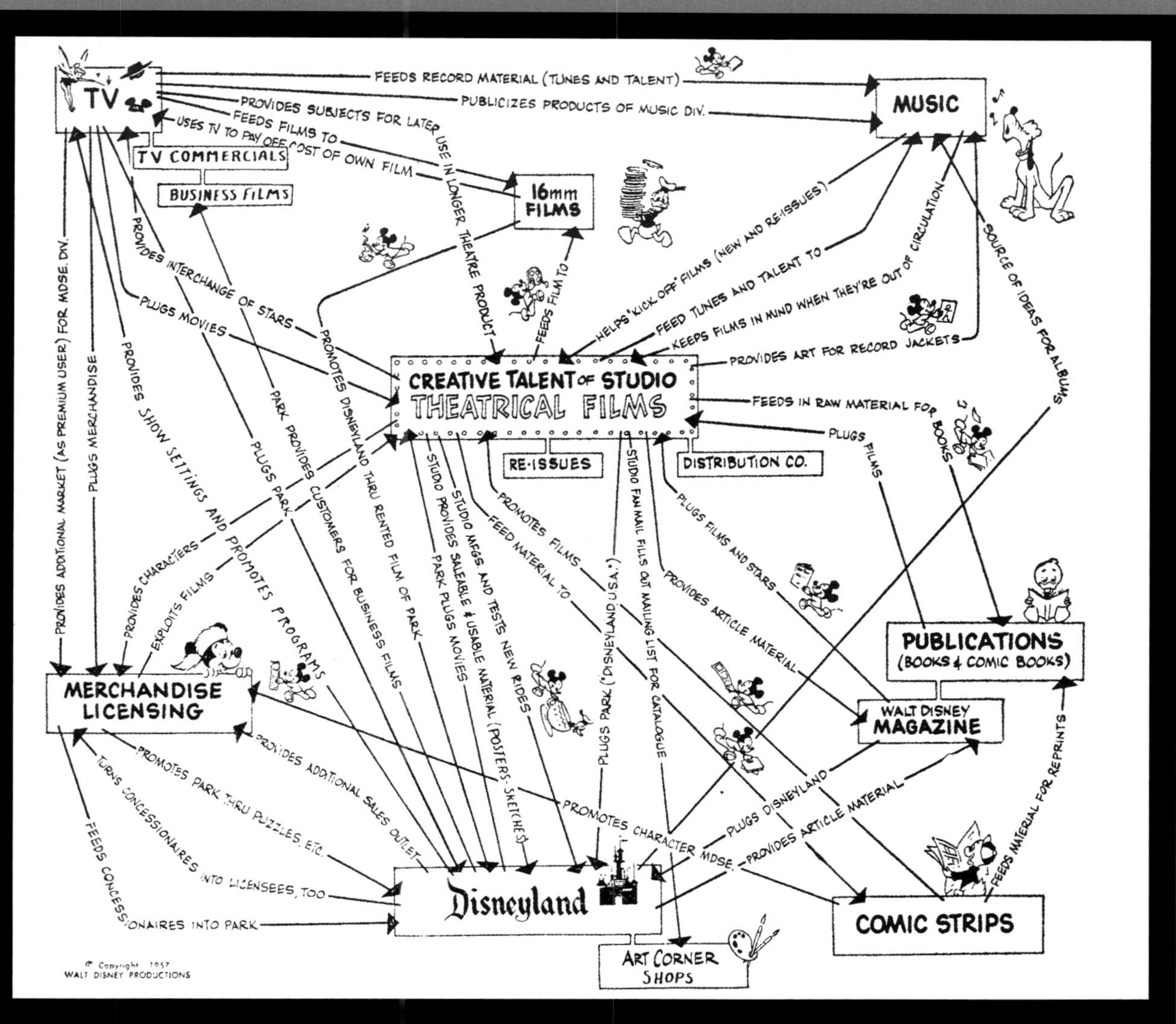
TV
MUSIC
FEEDS RECORD MATERIAL (TUNES AND TALENT)
PUBLICIZES PRODUCTS OF MUSIC DIV.
PROVIDES SUBJECTS FOR LATER USE IN LONGER THEATRE PRODUCT
FEEDS FILMS TO
USES TV TO PAY OFF COST OF OWN FILM
TV COMMERCIALS
BUSINESS FILMS
16mm FILMS
FEEDS FILM TO
PROVIDES INTERCHANGE OF STARS
PLUGS MOVIES
PROVIDES ADDITIONAL MARKET (AS PREMIUM USER) FOR MDSE. DIV.
PLUGS MERCHANDISE
PROVIDES SHOW SETTINGS AND PROMOTES PROGRAMS
PLUGS PARK
PARK PROVIDES CUSTOMERS FOR BUSINESS FILMS
PROMOTES DISNEYLAND THRU RENTED FILM OF PARK
HELPS "KICK OFF" FILMS (NEW AND RE-ISSUES)
FEED TUNES AND TALENT TO
KEEPS FILMS IN MIND WHEN THEY'RE OUT OF CIRCULATION
SOURCE OF IDEAS FOR ALBUMS
PROVIDES ART FOR RECORD JACKETS
CREATIVE TALENT OF STUDIO
THEATRICAL FILMS
FEEDS IN RAW MATERIAL FOR BOOKS
PLUGS FILMS
RE-ISSUES
DISTRIBUTION CO.
STUDIO PROVIDES SALEABLE & USABLE MATERIAL (POSTERS-SKETCHES)
STUDIO MFGS AND TESTS NEW RIDES
PARK PLUGS MOVIES
FEED MATERIAL TO
PROMOTES FILMS
PLUGS PARK ("DISNEYLAND U.S.A.")
STUDIO FAN MAIL FILLS OUT MAILING LIST FOR CATALOGUE
PROVIDES ARTICLE MATERIAL
PLUGS FILMS AND STARS
PUBLICATIONS
(BOOKS & COMIC BOOKS)
WALT DISNEY MAGAZINE
PROVIDES CHARACTERS
EXPLOITS FILMS
MERCHANDISE LICENSING
PROVIDES ADDITIONAL SALES OUTLET
PROMOTES PARK THRU PUZZLES, ETC.
TURNS CONCESSIONAIRES INTO LICENSEES, TOO
FEEDS CONCESSIONAIRES INTO PARK
PROMOTES CHARACTER MDSE.
PLUGS DISNEYLAND
PROVIDES ARTICLE MATERIAL
FEEDS MATERIAL FOR REPRINTS
Disneyland
COMIC STRIPS
ART CORNER SHOPS
© Copyright 1957
WALT DISNEY PRODUCTIONS

MAP YOUR RESOURCES

Identify what your firm does well (intangible resources) and the unique assets it has (tangible resources).

OVERVIEW

Mapping your resources will determine strengths and capabilities of your firm, which determines your competitive advantage.

TIME NEEDED

60–90 minutes

MATERIALS

Markers, sticky notes, whiteboard, Resource Cards

STEPS

1. Print out or download a set of Resource Cards. Create six columns on the whiteboard with headers: Resources, Value, Rarity, Imitability, Organization (VRIO), and Competitive Advantage. Below the Resources column label Tangible for the top half and Intangible for the bottom half.
2. Have participants ideate the firm's tangible and intangible resources, writing one resource per sticky note. Use the list of items on the Resource Cards as inspiration. A tangible resource is a physical item that you can touch. An intangible resource cannot be detected by any of the human senses. Post the resources in the Resources column.
3. Now review each resource and mark an X in each of the VRIO columns indicating the qualities of the resource:
   - Valuable: Value is determined if a resource adds value by enabling the firm to exploit opportunities or defend against threats.
   - Rare: Very few companies could acquire rare resources.
   - Inimitable: Another organization doesn't have the resource, or can't imitate, buy, or substitute it at a reasonable price.
   - Organized: Your firm is organized to capture the value from the resource.
4. Look across all four dimensions of each Resource row. Move from left to right, starting with Valuable. If the resource is not valuable, it could be outsourced, as it brings no value to the firm. If the resource is not rare, the firm is not worse than its competition. If the resource is not costly or difficult to imitate, other companies may try to imitate it in the future, and the firm does not have a competitive advantage. If you are not able to organize your company to capture value, the resource becomes expensive for the firm. Finally, if your firm can manage the resource advantages and is well-organized for execution, you have a significant competitive advantage.
5. Review your assessment. Look at the strength of your resources: what could be outsourced, changed, or improved? Synthesize your key insights and craft them into succinct, memorable full-sentence statements. Take photos of all whiteboards.

Additional activity resources and templates can be found at www.theartofopportunity.net.

SOURCE: Based upon the VRIO Framework, originally developed by Barney, J. B. (1991) in his work, *"Firm Resources and Sustained Competitive Advantage."*

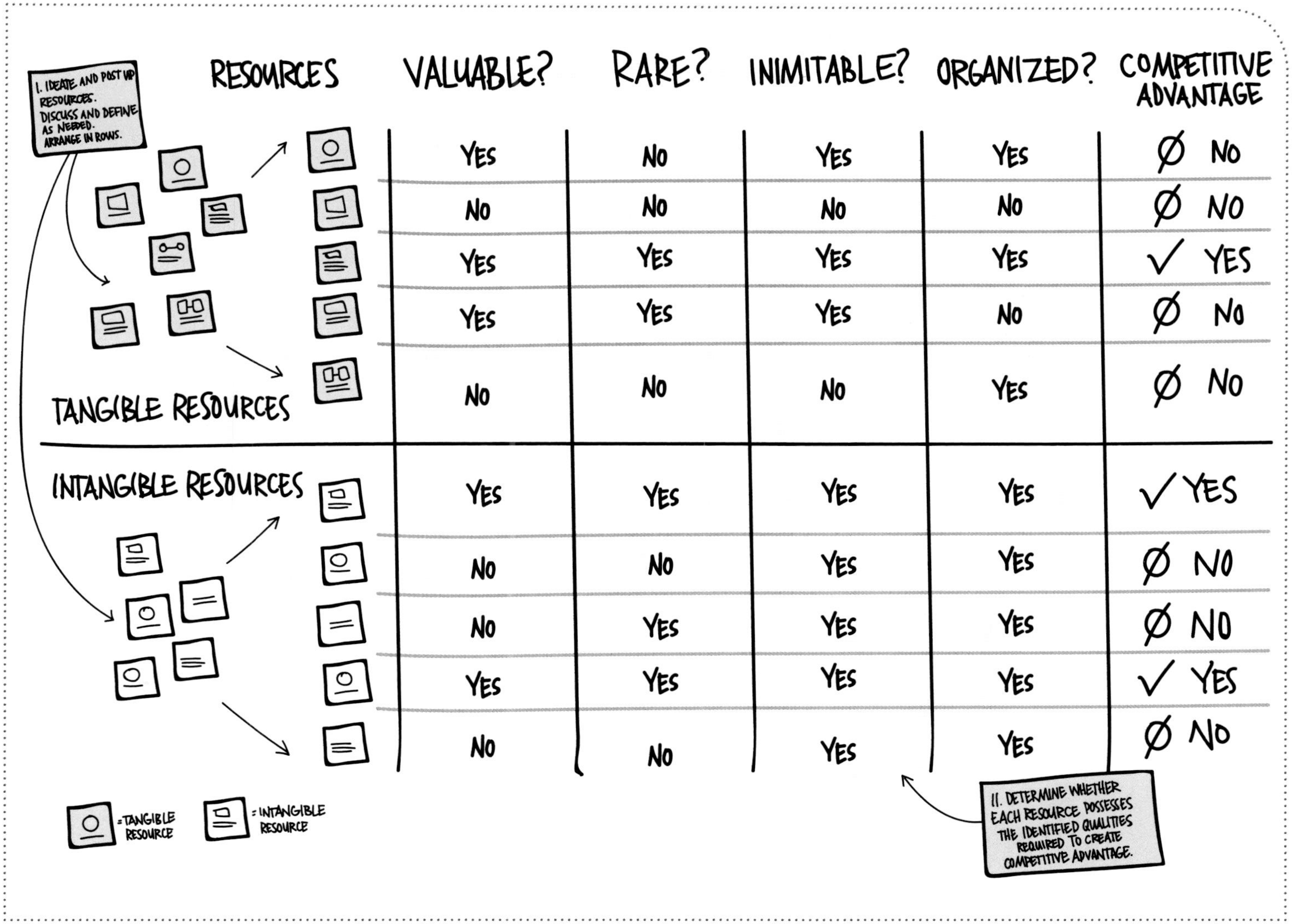

RESOURCES	VALUABLE?	RARE?	INIMITABLE?	ORGANIZED?	COMPETITIVE ADVANTAGE
TANGIBLE RESOURCES					
	YES	NO	YES	YES	Ø NO
	NO	NO	NO	NO	Ø NO
	YES	YES	YES	YES	✓ YES
	YES	YES	YES	NO	Ø NO
	NO	NO	NO	YES	Ø NO
INTANGIBLE RESOURCES					
	YES	YES	YES	YES	✓ YES
	NO	NO	YES	YES	Ø NO
	NO	YES	YES	YES	Ø NO
	YES	YES	YES	YES	✓ YES
	NO	NO	YES	YES	Ø NO

I. IDEATE AND POST UP RESOURCES. DISCUSS AND DEFINE AS NEEDED. ARRANGE IN ROWS.

II. DETERMINE WHETHER EACH RESOURCE POSSESSES THE IDENTIFIED QUALITIES REQUIRED TO CREATE COMPETITIVE ADVANTAGE.

= TANGIBLE RESOURCE

= INTANGIBLE RESOURCE

MAP YOUR ECOSYSTEM

Create a basic ecosystem diagram using three elements: Stakeholder roles (nodes), activities (links), and deliverables (another node form).

OVERVIEW

Visualize your business ecosystem to understand the holistic picture of the stakeholders (roles), deliverables, and the interrelationship of activities.

TIME NEEDED

60–90 minutes

MATERIALS

Markers, sticky notes, blank wall or whiteboard

STEPS

1. Give each participant sticky notes and markers. Have the team ideate your set of stakeholder roles/nodes. These are all the stakeholder participants: firm leaders and departments, suppliers, contract service providers, customers, and other organizations. Only write one role per sticky note. Think about who provides what resources or deliverables.
2. Next, group stakeholders in affinity sets of attributes, such as suppliers, distributors, partners, customers or clients, contractors, and influencers (media, government/regulatory, or trade associations).
3. Now use the tangible and intangible resources you mapped in the "Map Your Resources" activity as inspiration to develop a set of deliverables that create value (tangible or intangible) between two parties. Use different colors for tangible and intangible deliverables.
4. Next, draw lines with arrows to show the direction of the activity transaction connecting the roles. The lines of transactions represent the value exchanged between roles in an ecosystem. Examples of a transaction are: payment, documents, equipment, contracts, schedules, workbooks, advice, security, feedback, approval, criticism, or assurance.
5. Review the final ecosystem and validate all the roles, deliverables, and transactions with the team. Consider getting additional input and feedback from other people in your firm. Synthesize your key insights and craft them into succinct, memorable full-sentence statements. Take photos of all whiteboards.

Additional activity resources and templates can be found at www.theartofopportunity.net.

SOURCE: This activity was created by the *Art of Opportunity* authors, and inspired by the *Value Network Analysis*, developed by Verna Allee.

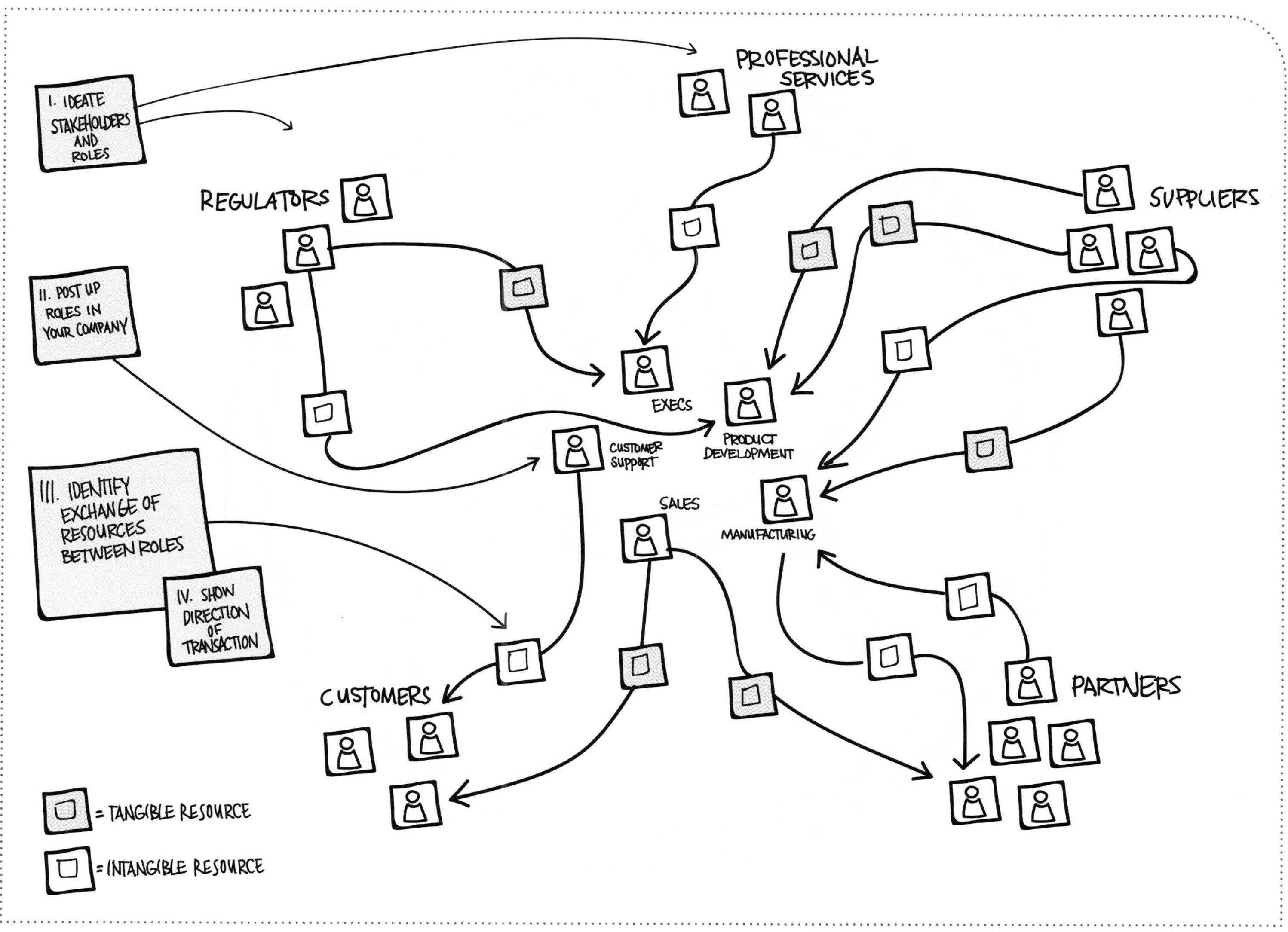
I. IDEATE STAKEHOLDERS AND ROLES
II. POST UP ROLES IN YOUR COMPANY
III. IDENTIFY EXCHANGE OF RESOURCES BETWEEN ROLES
IV. SHOW DIRECTION OF TRANSACTION
PROFESSIONAL SERVICES
REGULATORS
SUPPLIERS
EXECS
PRODUCT DEVELOPMENT
CUSTOMER SUPPORT
SALES
MANUFACTURING
CUSTOMERS
PARTNERS
= TANGIBLE RESOURCE
= INTANGIBLE RESOURCE

FRAME YOUR GROWTH INITIATIVE

HAVING EXPLORED YOUR CUSTOMER and noncustomer and gained a better understanding of your firm by mapping your resources and visualizing your ecosystem, it's time to make some decisions. What kind of growth are you looking for? What is within the boundaries of what is imaginable and what is not? How bold are you willing to be? Framing your growth initiative will allow you to stay focused when crafting your strategy. If you are the CEO or boss, and would like to task a team with the initiative, a growth brief will give them a bit more guidance about what to look for, and will help you to communicate your expectations clearly. In the case where you are the executive who has been tasked by your senior executive to develop a new growth strategy, the growth brief will help you to clarify expectations and make sure your efforts are focused on what the company wants to achieve.

This does not mean that throughout the process you cannot look elsewhere or go outside your comfort zone. The brief nevertheless enables you to compare the various opportunities and strategies to your original goals.

We don't want to dismiss several valid and worthwhile traditional growth models that can, and should be, applied for certain situations to meet specified goals. However, we are not focused on traditional forms of growth that are often a result of organizational process optimization, tweaking product features, or bolting on a new company. Before we describe the types of growth we are addressing in *The Art of Opportunity*, let's tell you about two specific types of growth we are not addressing.

WE ARE NOT FOCUSED ON THESE TWO GROWTH TYPES:

1. **Selling More of the Same:**
   A business that chooses to grow organically must expand using existing resources and processes to accommodate its growth. The idea is basically to try to increase volume by producing and selling more of the same. Market penetration occurs when a firm enters a market in which its current products already exist or its services are provided, allowing the business to go head-to-head with incumbents in the market.

2. **Growth through Mergers and Acquisitions:**
   Mergers and acquisitions are often undertaken to simply increase the size of a firm (think about Hewlett-Packard buying Compaq, or Marriott buying Starwood Hotels) and/or move up in the food chain of its industry. Some companies buy others to add capabilities to their portfolio that they don't possess, whereas others want to add product lines outside of their current core businesses to diversify (think Microsoft buying Nokia's smartphone products).

ART OF OPPORTUNITY GROWTH TYPES

WE SEE THREE TYPES OF GROWTH you want to address using *The Art of Opportunity*, which we are calling evolutionary growth, adjacent growth, and breakthrough growth. In what follows, we are focused on breakthrough growth, while our approach provides thoughtful and practical insights into evolutionary and adjacent growth.

EVOLUTIONARY GROWTH

Evolutionary growth designates a growth type that is closest to your core. You evolve your existing offering to remove hurdles to satisfaction and barriers to consumption. This might mean enhancing the design of your products to make them more user-friendly or upgrading your services.

Consider the example of the train company upgrading its in-train services and reducing travel time between two cities to attract business customers, or a restaurant chain refurbishing its dining rooms to make them more comfortable. McDonald's, for example, invests in improving the quality of its food and upgrading its restaurants. Amazon makes the shopping experience easier and more convenient by reducing the risk of online purchases through such features as "Look inside" and offering free returns, no questions asked. Cardinal Health made the planning, ordering, and delivery of surgical tool kits easier and more convenient.

Evolutionary growth is usually the easiest, while also being the least risky. On the other hand, it is often the least sustainable one, and while it can reduce barriers to consumption and hurdles to satisfaction, it can often be copied easily by competitors.

ADJACENT GROWTH

Adjacent growth designates when a company brings new products or services to market that are not too far away from its core business and actually are closely related to and complement existing offerings. Adjacent growth often means expanding your offering to cover additional steps in the customer experience or offering other similar products.

Consider again the example of the train company. While evolutionary growth would have it upgrade the trains and the existing products, services, and customer experience, adjacent growth introduces new products and services around the existing ones. For the train company, this could mean offering transfers to and from the railway station, for example, or offering luggage transportation and handling services. McDonald's went from fast food to introducing McCafé as an adjacent offering while staying close to its core. Amazon added further products to its platform, going beyond selling books to offering a broad

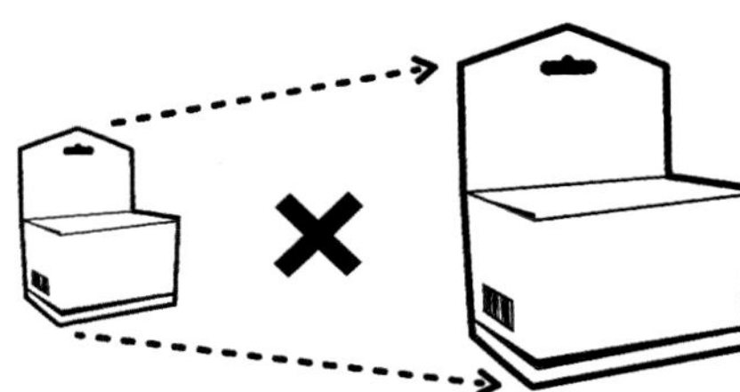

EVOLUTIONARY GROWTH
Improve your current offering.

ADJACENT GROWTH
Complement your current offering.

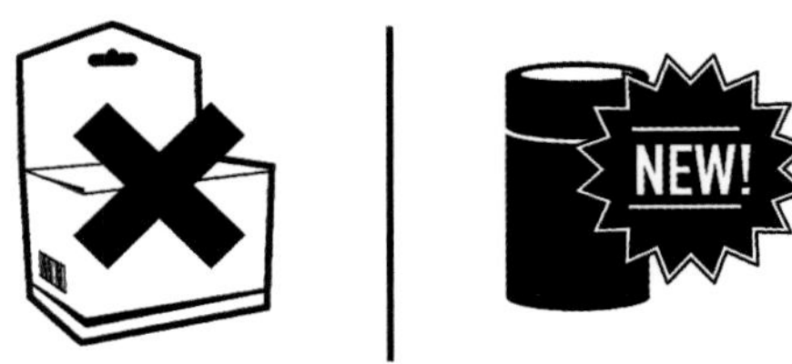

BREAKTHROUGH GROWTH
Develop an entire new offering.

ART OF OPPORTUNITY GROWTH TYPES

spectrum of products ranging from home and garden tools to beauty and health care products, sports and outdoors items, toys, clothing, and automotive and industrial products, among others.

Adjacent growth bears a little more risk than evolutionary growth, as you are venturing into slightly new territory. Yet, as you are staying close to your core, nevertheless the risk is manageable. Adjacent growth as we have outlined it here usually focuses on extending your offering to address additional steps in the customer experience, reducing hurdles to satisfaction and hence making the overall journey more pleasant for the customer.

BREAKTHROUGH GROWTH

Breakthrough growth is the type of growth that breaks through and goes well beyond the limits of your current business. Breakthrough growth often entails not only the development and launch of a completely new strategy to market an offering outside of your company's existing business definition, but also the design of a new business model and/or revenue model as part of the new strategy.

As an example, Google's core business is search. However, Google is making new growth bets on self-driving cars, optical head-mounted display glasses, and Android Wear for watches and wearables. Amazon went beyond offering typical retail products on its platform to offering Amazon Web Services; providing payment, fulfillment, and logistics services; developing and producing the Kindle; and selling groceries through AmazonFresh.

Breakthrough growth is obviously not only the most difficult, but also the most risky type to achieve. Yet it also bears the highest rewards, if successful. And we will show you how the risk can be managed!

Which type of growth are you aiming at? Let's define your type of growth initiative. Each growth model is appropriate for specific situations. There is not one prescribed model type and, in fact, you may benefit from combining them in order to adapt to your firm's individual situation.

Clarifying your objectives and the type of growth you are aiming at will enable you to focus your subsequent strategy efforts, provide your team with guidance, and avoid pursuing opportunities your company might not be comfortable with at present.

THE BREAKTHROUGH QUESTION

If you choose to go for **breakthrough growth** and strategic innovation in the truest sense, we suggest summing up your opportunity in what we call a **Breakthrough Question**. Breakthrough Questions ask about solving seemingly impossible challenges and tensions. Breakthrough Questions go beyond the usual "What if we could . . .?" inquiries, which, more often than not, are answers disguised as questions. At this stage, you shouldn't jump to conclusions and answers just yet. The better the Breakthrough Question, the more likely you are to come up with an innovative strategy. Faulty questions yield faulty strategies.

Consider these examples:

- How can we increase value for our customers, while lowering cost for our company?
- How can we give away our product for free, while increasing our profit?
- How can we make our product available to the least profitable customers, while keeping our margins?
- How can we make our product available to customers who cannot afford it at the moment, while keeping our margins?
- How can we leverage our key assets outside of existing customer segments?
- How can we use our existing assets to address the needs of our noncustomers?
- How can we expand our network infrastructure without investing in it?
- How can we decrease our lead time and increase delivery speed without increasing inventory or logistics cost?

FRAME YOUR GROWTH OPPORTUNITY

HAVING DEFINED YOUR GROWTH OBJECTIVES, the final step is to align your identified growth opportunities with your objectives. The opportunities you discovered will probably fall into one of the three types of growth categories outlined earlier.

DECISION MAKING

We all make decisions of varying importance every day, such as what to wear for a big event, where to hold a company off-site meeting, or when to buy a new car. We're pros at decision making, right? So the idea that you should apply a more sophisticated approach may seem unusual. The reality is that most people are not well-trained or practiced in effective decision making. A number of different strategies and decision paths can be applied to finding or selecting which opportunities you will explore.

So what makes a good opportunity? The market size, leveraging competencies, a completely new offering, or meeting an underserved customer need? The answer is potentially all of these criteria. In all cases, we establish our goals and objectives, develop a set of opportunity alternatives, and then apply criteria to evaluate them. Further, our objective is to select the opportunities with the highest probability of effectiveness or success, as well as those that best fit our goals and objectives.

Decision making is seldom a black-and-white linear process. In fact, for businesses, nonlinear, iterative, and recursive thinking is best applied for a number of reasons, the least of which is reducing turnaround, costs, uncertainty, and doubt as time, stakeholders, and other influences come to bear. These are some of the key benefits of the active iteration principle of business design thinking, which is outlined in Chapter 5.

NONCOMPENSATORY AND COMPENSATORY DECISION MAKING

NONCOMPENSATORY AND COMPENSATORY

Initial metric for performance my be effective in the short-term, but overlook long term advantages.

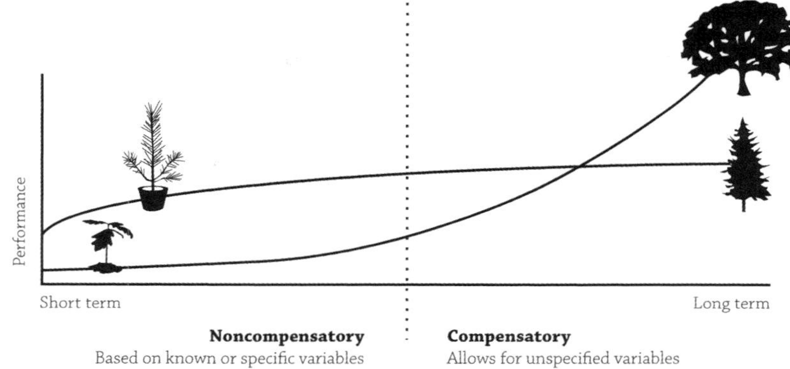

KINDS OF DECISION-MAKING STRATEGIES

For years, scientists have developed decision-making models. Many have theoretical and off-putting names, such as the linear model, the additive difference model, the ideal point model, the conjunctive model, the disjunctive model, the lexicographic model, and the elimination-by-aspects model.

Despite their formidable names, most of these strategies are actually quite common. Decision making can be viewed in two primary groups, compensatory and noncompensatory. A compensatory strategy confronts the conflict in the choice situation. The noncompensatory strategies avoid the conflict in the choice situation. Compensatory strategies allow you to trade off a low score on one criterion for a high score on a different criterion, whereas noncompensatory strategies do not allow such trade-offs.

We believe outcomes are improved with variety, so we suggest applying more than one strategy to evaluate your opportunity alternatives and make a choice. To reduce the amount of information you need to process, we suggest using a noncompensatory strategy to thin out your set of alternative opportunities. Then use a compensatory strategy to conduct a careful analysis of the remaining alternatives.

Together, these strategies let you structure the relevant information available, make efficient use of your imperfect information-processing capabilities, and reach conclusions that are more sound than simple gut instinct alone. And by applying these strategies to choose among alternatives, you will be able to make more effective use of them later on and have defensible decisions that can be justified to others based on the information you have available.

DECISION-MAKING PROCESS

Your goal is to advance your best opportunity idea(s) to the next phase of crafting a strategy. As such, your decision is very important. You will want to apply a process that gets you to a decision swiftly and has minimal risk so that you can iterate and learn quickly.

While you always want consensus from your diverse stakeholders, group decision making could slow you down and may create a watered-down idea by giving every voice equal say. There is a fine balance between empowering a team or individual to use best judgment and using group decision making to gain full consensus.

A reasonable balance of methods is to apply consent-based decision making versus simple consensus. Using a consensus approach, everyone has an equal say and the majority rules. In consent-based decision making, those with dissenting ideas have the opportunity to present their reasons why, and a leader or the group reconsiders before rendering a decision and citing the logic for the choice. Your objective is to make a decision that has no valid objections, while ensuring that dissenting participants have been heard and are willing to live with the choice.

It is important to acknowledge that there is no perfect answer. But through effective team decision making, you have the opportunity to empower teammates, get to answers more quickly, and therefore take action sooner, so you can learn and evolve your opportunity and its strategy.

MOST PEOPLE ARE

NOT WELL-TRAINED

OR PRACTICED

IN EFFECTIVE

DECISION MAKING.

CREATE YOUR GROWTH INITIATIVE BRIEF

The Growth Initiative Brief activity builds a framework to focus your efforts and evaluate your progress.

OVERVIEW

Clarifying your growth initiative objectives, and the type of growth you hope to achieve, enables you to focus your subsequent strategy efforts, and provides your team with alignment, clarity, and guidance.

TIME NEEDED

60–120 minutes

MATERIALS

Growth Initiative Brief Template, whiteboard

STEPS

1. Print out or download a Growth Initiative Brief template. Name one team member as the scribe. Write the headers for each category on the whiteboard: Goal, Why, Parameters, Growth Team, Available Resources, Sponsors, Decision Process.
2. Have a discussion around each topic. Use sticky notes for initial ideas, so everyone's voice can be heard. Start with your **Goal:** What would you like to accomplish? Think with the end in mind and work backward. Is your goal determined by earnings, revenue, profit, or market share, or to be #1 in the industry?
3. Now move to the **"Why?"** Why do you want or need to significantly grow your business or start a new business? What problem are you solving? What is your strategic intent? Diversification, expansion, to build a broader network of suppliers, industry disruption?
4. Next move to **"Parameters."** Are parameters already established? Are you required to stay within your current industry? What is your timing? Do you have to demonstrate revenue and/or profit within one year? 3 years? 5 years?
5. Identify the members of your **Growth Team**. What is your team size? Do you have a diverse mix of members? Do you have optimal representation from internal functions within the firm? Should you include outside experts, customers, or vendors?
6. Finish up with the last three brief topics. For **Available Resources**, do you have any seed funds? What human resources do you have? Do you have access to outside resources? What technology is available? Name your **Sponsors**, both formal and informal, i.e. the firm's board, key leaders, or a division of the company. Finally, outline the **Decision Process**. Who is involved and what are their roles (e.g., fund, approve, influence)?
7. Complete the brief, and share with key stakeholders for input and feedback. Take photos of all whiteboards.

Additional activity resources and templates can be found at www.theartofopportunity.net.

SOURCE: *The Art of Opportunity* authors

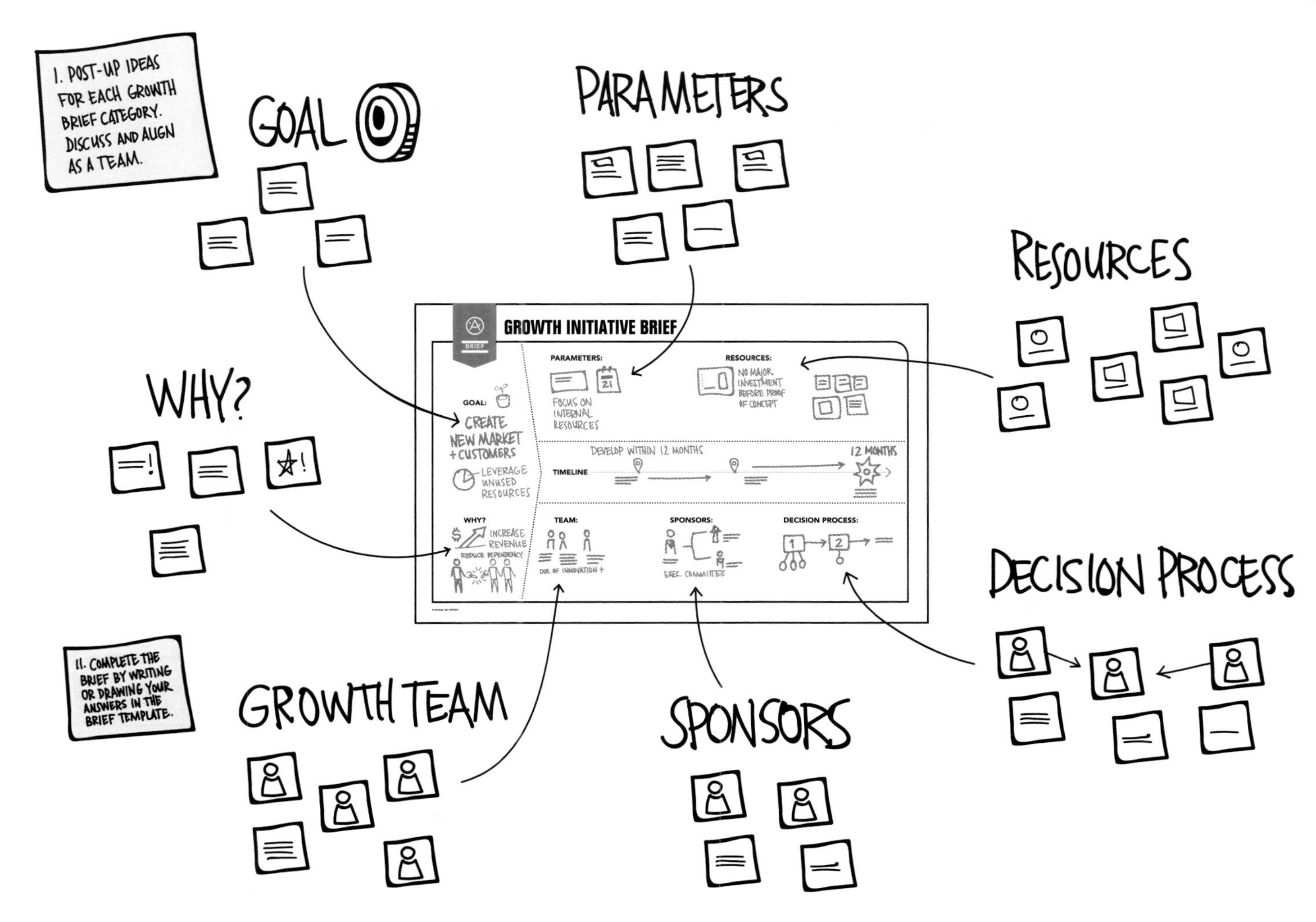
I. POST-UP IDEAS FOR EACH GROWTH BRIEF CATEGORY. DISCUSS AND ALIGN AS A TEAM.
GOAL
PARAMETERS
RESOURCES
WHY?
GROWTH INITIATIVE BRIEF
GOAL:
CREATE NEW MARKET + CUSTOMERS
LEVERAGE UNUSED RESOURCES
PARAMETERS:
FOCUS ON INTERNAL RESOURCES
RESOURCES:
NO MAJOR INVESTMENT BEFORE PROOF OF CONCEPT
TIMELINE
DEVELOP WITHIN 12 MONTHS
12 MONTHS
WHY?
INCREASE REVENUE
REDUCE DEPENDENCY
TEAM:
DIR. OF INNOVATION +
SPONSORS:
EXEC. COMMITTEE
DECISION PROCESS:
DECISION PROCESS
II. COMPLETE THE BRIEF BY WRITING OR DRAWING YOUR ANSWERS IN THE BRIEF TEMPLATE.
GROWTH TEAM
SPONSORS

EXPLORE YOUR OPPORTUNITY INSIGHTS

Use visualization to parse through nuggets of insight to land on a set of opportunity statements.

OVERVIEW

Visualizing patterns, themes, and clusters from among your set of insights helps you get organized to find and develop clear opportunity statements.

TIME NEEDED

½ day in preparation; 120 –180 minutes in workshop

MATERIALS

Markers, sticky notes, and whiteboard

STEPS

1. This is a major effort and critical step to finding your opportunity. Prepare by reviewing the goals and objectives in the Growth Initiative Brief that you created. Select the key goals you outlined that you will use to evaluate the suitability of your opportunity for meeting your objectives.
2. Gather all your supporting observation material, and your customer and noncustomer persona maps. Post up all your insight statements from your field observation, the assessment of your firm's resources, your customer and noncustomer journey map, and visualized ecosystem.
3. Now review the materials for complementary insights. Look for key points that seem strong or weak, have key dependencies, or are mutually exclusive. Look for patterns, themes, and relationships between your field insights, personas, and your company ecosystem. This step can be messy, but seeing your opportunities up on a whiteboard will make it easier to land on key insights. Feel free to post small notes and descriptions next to insights.
4. Organize the clusters of insights in relation to each other. Next you will name your possible opportunities. Naming your opportunity is a restatement of problems that you find in the unmet needs or expectations of your customer or noncustomer (basically, the job to be done). For example, "our customer needs a faster, less expensive alternative to public transportation." Generate as many opportunity statements as possible, but naming your opportunity is not finding a solution. In fact, identifying an unmet need or expectation may spawn multiple opportunities.
5. Select the set of opportunity statements that most spark your imagination and have the greatest likelihood of success, when compared to your goals and objectives. Allow for wildcards, for the craziest opportunities. Don't just settle on the ones you consider being the easiest to seize. Each opportunity will have a named customer, based on your research. Take photos of all whiteboards.

Additional activity resources and templates can be found at www.theartofopportunity.net.

SOURCE: *The Art of Opportunity* authors

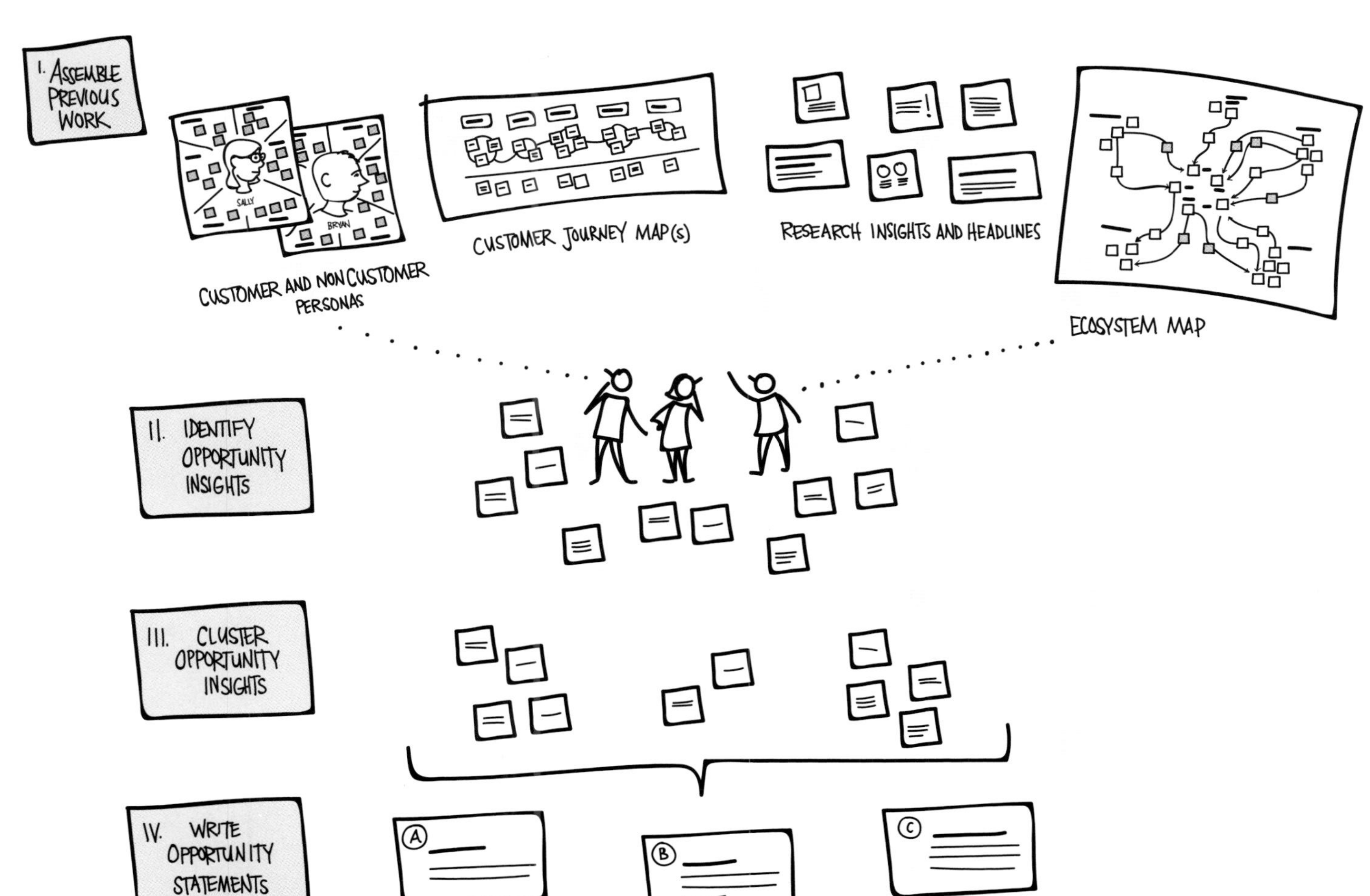
I. ASSEMBLE PREVIOUS WORK
SALLY
BRYAN
CUSTOMER AND NON CUSTOMER PERSONAS
CUSTOMER JOURNEY MAP(S)
RESEARCH INSIGHTS AND HEADLINES
ECOSYSTEM MAP
II. IDENTIFY OPPORTUNITY INSIGHTS
III. CLUSTER OPPORTUNITY INSIGHTS
IV. WRITE OPPORTUNITY STATEMENTS
A
B
C

SELECT YOUR OPPORTUNITY

Select the opportunity, or opportunities, and create the breakthrough question(s) you will use in crafting your strategy phase.

OVERVIEW

Apply a variety of decision-making processes to set the opportunities you believe have the greatest chance of success, or are the most innovative ones, and/or the most challenging ones to be used in the next phase of crafting your strategy.

TIME NEEDED

45–60 minutes

MATERIALS

Markers, sticky notes, and whiteboard

STEPS

1. Now that you have identified a number of opportunities, you will work with your team to assign value to see how well each opportunity statement meets the criteria of feasibility, desirability, and viability. Begin by giving your participants sticky notes, markers, and a copy of your Growth Initiative Brief.
2. Create a grid on the whiteboard and write each of your opportunity statements at the top of each column. Then label 4 rows: Feasibility, Desirability, Viability, and Totals.
3. Have each team member privately assign a numerical value to each cell, using a 1–5 rating, with 5 representing the optimal degree to which they believe each opportunity statement has the greatest feasibility (can we do this?), desirability (do users want it?), and viability (should we do this?). Have participants post their ratings in each cell of the grid.
4. Have the scribe tally the total combined score and create the average score for each cell, and in the totals row, put the sum of all total and average cell tallies.
5. Now, have a team conversation to ensure alignment and consent. If there is concern or disagreement, have a discussion to decide if you change your selection.
6. As a group, draft breakthrough questions for your top two opportunities using the Breakthrough Question template (download at www.theartofopportunity.net).

Additional activity resources and templates can be found at www.theartofopportunity.net.

SOURCE: *The Art of Opportunity* authors

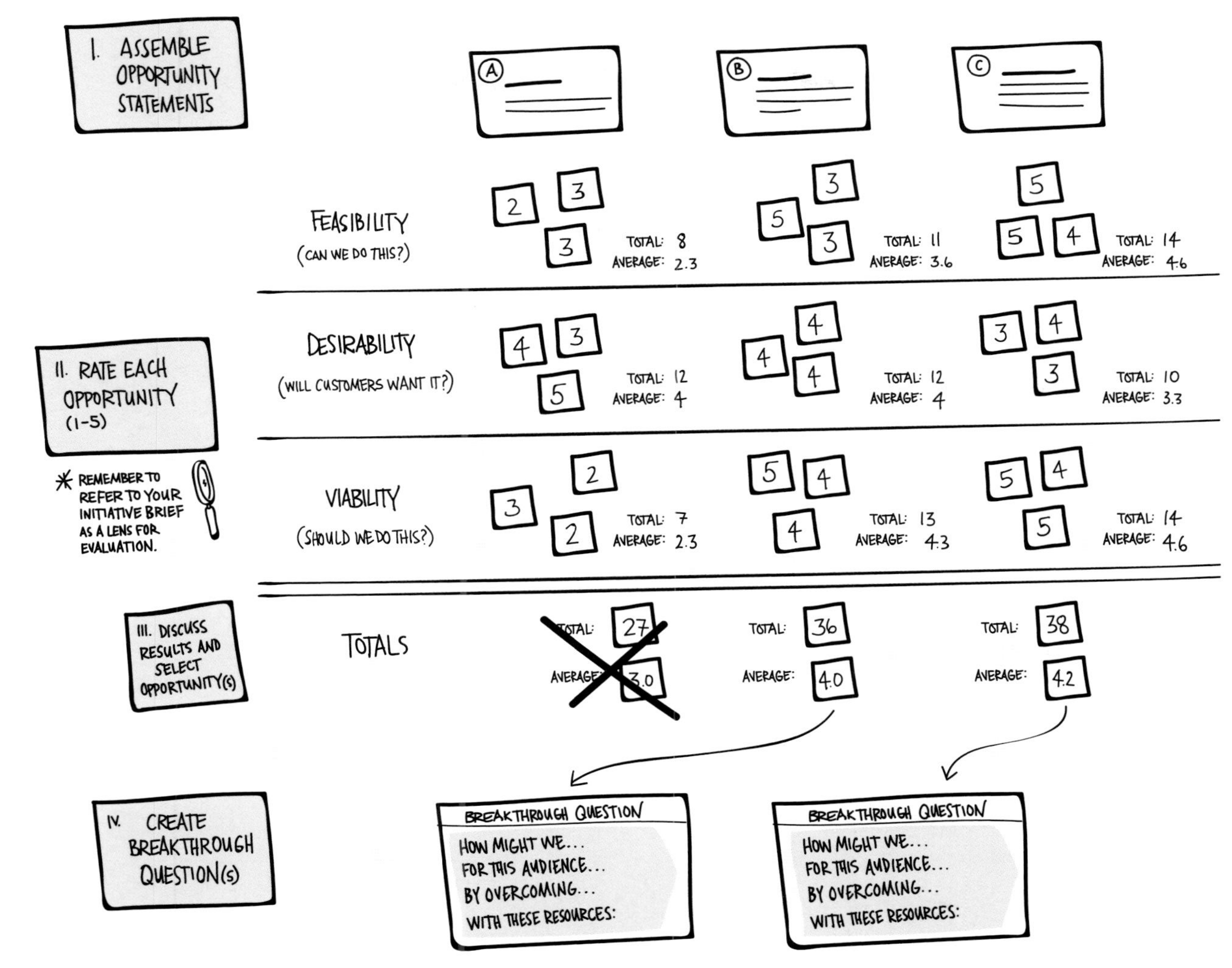

I. ASSEMBLE OPPORTUNITY STATEMENTS
A
B
C
FEASIBILITY (CAN WE DO THIS?)
2 3 3
TOTAL: 8
AVERAGE: 2.3
3 5 3
TOTAL: 11
AVERAGE: 3.6
5 5 4
TOTAL: 14
AVERAGE: 4.6
II. RATE EACH OPPORTUNITY (1-5)
* REMEMBER TO REFER TO YOUR INITIATIVE BRIEF AS A LENS FOR EVALUATION.
DESIRABILITY (WILL CUSTOMERS WANT IT?)
4 3 5
TOTAL: 12
AVERAGE: 4
4 4 4
TOTAL: 12
AVERAGE: 4
3 4 3
TOTAL: 10
AVERAGE: 3.3
VIABILITY (SHOULD WE DO THIS?)
3 2 2
TOTAL: 7
AVERAGE: 2.3
5 4 4
TOTAL: 13
AVERAGE: 4.3
5 4 5
TOTAL: 14
AVERAGE: 4.6
III. DISCUSS RESULTS AND SELECT OPPORTUNITY(S)
TOTALS
TOTAL: 27
AVERAGE: 3.0
TOTAL: 36
AVERAGE: 4.0
TOTAL: 38
AVERAGE: 4.2
IV. CREATE BREAKTHROUGH QUESTION(S)
BREAKTHROUGH QUESTION
HOW MIGHT WE...
FOR THIS AUDIENCE...
BY OVERCOMING...
WITH THESE RESOURCES:
BREAKTHROUGH QUESTION
HOW MIGHT WE...
FOR THIS AUDIENCE...
BY OVERCOMING...
WITH THESE RESOURCES:

VISUALIZE YOUR OPPORTUNITY

YOU'VE REACHED A MILESTONE. You have explored the segments of your customers and noncustomers, delved into their needs and expectations, and identified their barriers to consumption and hurdles to satisfaction. You've gone outside to observe and gather information about customers' journeys, and then framed up insights. Then you dove into your firm's ecosystem to mine more insights. Finally, you distilled this wealth of learning into a breakthrough opportunity question. Bravo!

You obviously need to share this excellent work with your boss and associates. How about a wildly compelling PowerPoint? NO! We've designed an Opportunity Report template so you can tell the story of this phase of your opportunity journey. Complete it with your teammates. It may take a few days of iteration. Leverage all your content and be creative. Add photos and drawings, and tell the story of how you landed on "where to play."

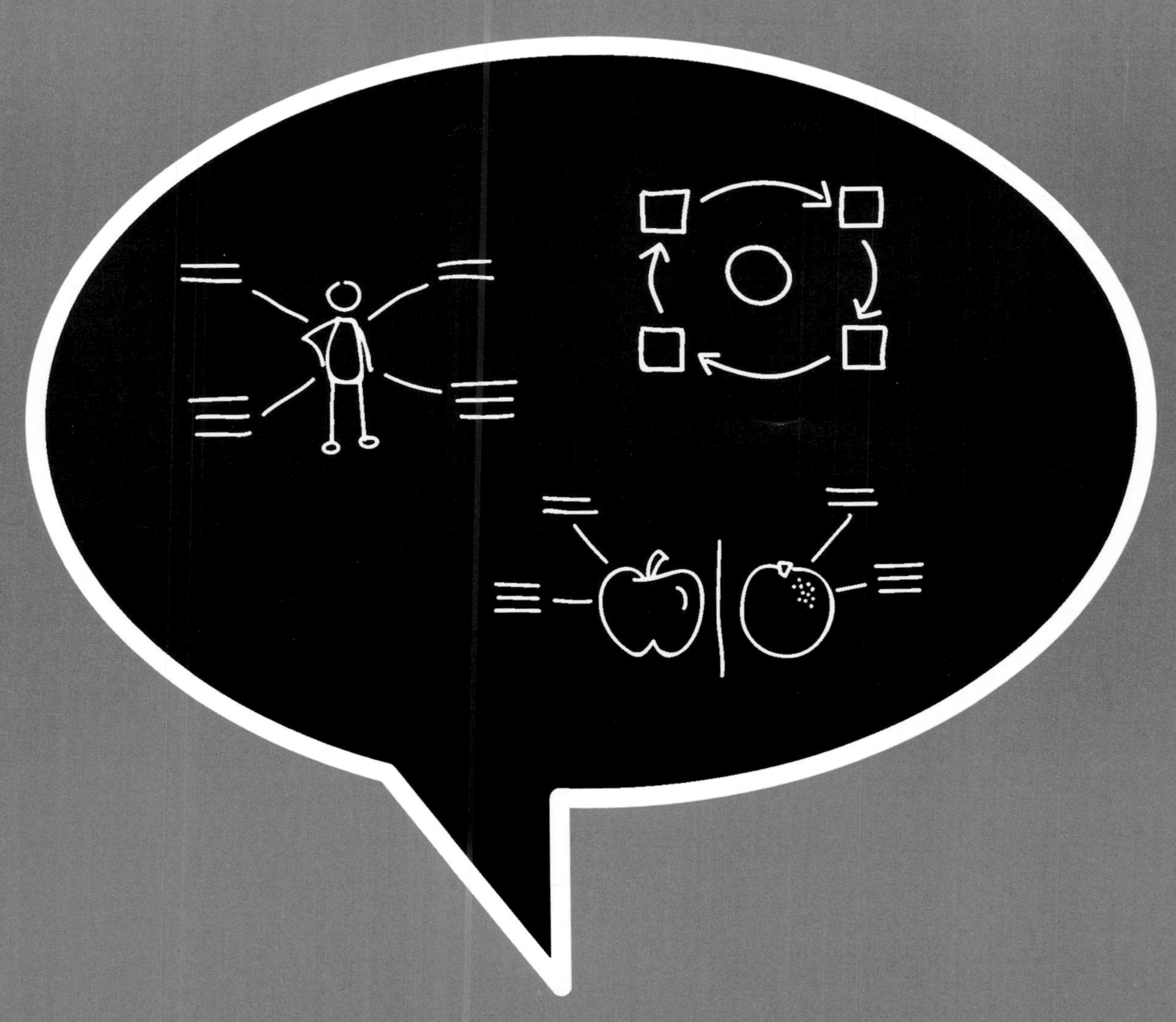

EMPLOY STORYTELLING AND VISUAL FRAMEWORKS

MOST PEOPLE WOULD AGREE that lengthy PowerPoint decks, Gantt charts, and copy-heavy memos are not engaging or effective ways to express new ideas. And while we recognize that not everyone is a designer, in our experience any attempt to build a cohesive visual story out of your content will be welcomed (or at least appreciated) because it's not more of the same—and isn't breaking out of "sameness" the point of your whole growth initiative? So, recognizing that not everyone is a master artist or storyteller, here are some tips for creating and building a compelling presentation:

- Begin by getting to the point. How simply can you tell your story? Can it be clear and concise with a simple storyline containing an introduction, a complicating situation, and a definitive resolution? Think in terms of headlines before adding colorful anecdotes or details.
- Add color by injecting classical story elements such as action, conflict, and transformation. Use characters and create tension by introducing problems that are not immediately solved, or may have an unusual resolution.
- Build a narrative that you feel comfortable presenting.

As humans we relate to stories about other people. Think about giving your audience a persona with whom they can sympathize. Leverage your persona map to identify a customer or noncustomer on whom to focus. Follow the process you used on your journey. Bring out the emotion in the customer's needs, for example, and present the human side of the barriers to be overcome. Demonstrate what it means for the customer when their need or desired experience is achieved.

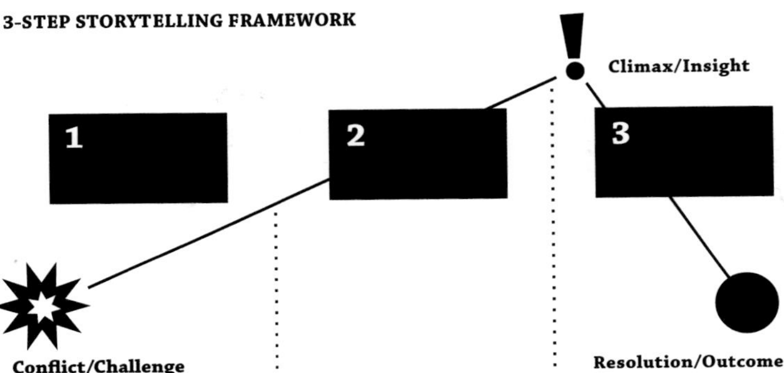

STEP 1
Establish characters, setting, and plot.

Questions to answer:
Who is this about?
What is the challenge?
Why should we care?
Where is this taking place?
What are the considerations?
What needs to be solved?

STEP 2
Identify obstacles, increase complexity, and build tension.

Questions to answer:
What are the complicating factors?

STEP 3
Reveal the moment of clarity, decision, and/or breakthrough learning.

Questions to answer:
What is the opportunity?
What is/are the key insight(s)?

NARRATIVE FLOW
Presenting your story in a chronological format mimics the classic "hero's journey." Reverse the direction of the narrative to present a "mystery" that carefully reveals how you arrived at your conclusions.

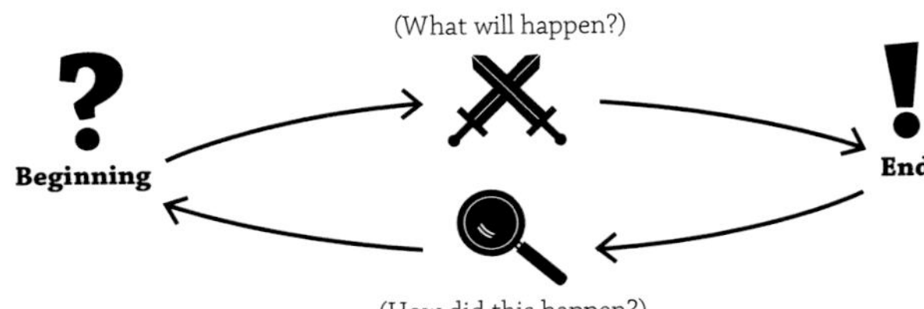

EMPLOY STORYTELLING AND VISUAL FRAMEWORKS

Even if you cannot draw, visual frameworks can help tell your story by providing readymade systems for organizing your information. Often, these frameworks can help explain what you did, how you did it, and why you did it in ways that your audience will intuitively grasp. Some common visual frameworks used by designers include:

- Relations (a network, matrix, or map)
- Groupings (tree, Venn, matrix, or map)
- Hierarchies (tree or Venn)
- Processes or sequences (showing flow, time, or steps)
- Locations (map)
- Quantities (bar or pie chart, time series)
- Comparisons (contrasting images or situations)

The use of storytelling and visual frameworks doesn't need to be complicated. For more inspiration or to learn in more detail, we suggest you explore the work of some of our favorite storytellers and designers.

DAN ROAM

- *The Back of the Napkin (Expanded Edition): Solving Problems and Selling Ideas with Pictures*
- *Blah Blah Blah: What To Do When Words Don't Work*

NANCY DUARTE

- *Slide:ology: The Art and Science of Creating Great Presentations*
- *Resonate: Present Visual Stories That Transform Audiences*
- *Illuminate: Ignite Change through Speeches, Stories, Ceremonies, and Symbols*

DAVE GRAY

- http://xplaner.com/visual-thinking-school/

SUNNI BROWN

- *The Doodle Revolution: Unlock the Power to Think Differently*

In her book, *The Doodle Revolution*, Sunni illustrates some visual frameworks that you can use to explain your story. Using her set of illustrations, you can see examples of system frameworks to describe a concept and show its parts—explaining what it is; process frameworks tell how something works and show its flow and/or sequence; and comparison frameworks to convince viewers of the merits of one concept over another adjacent.

VISUAL FRAMEWORKS like these from Sunni Brown can help you organize your information for more easily understandable presentations or reports.

SYSTEMS

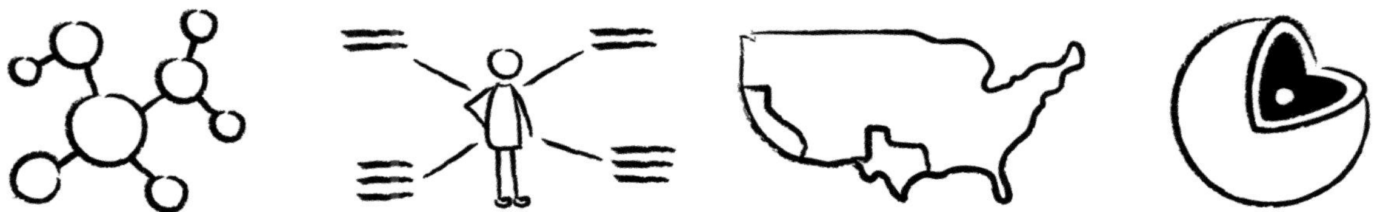

PROCESSES

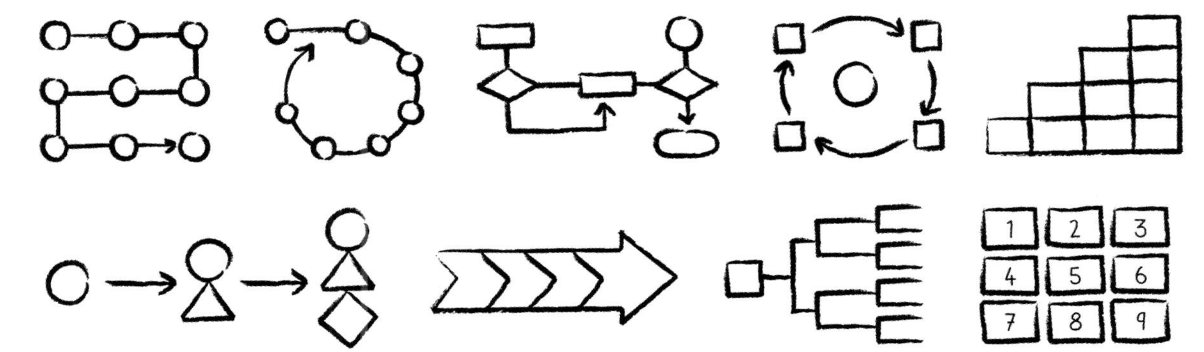

COMPARISONS

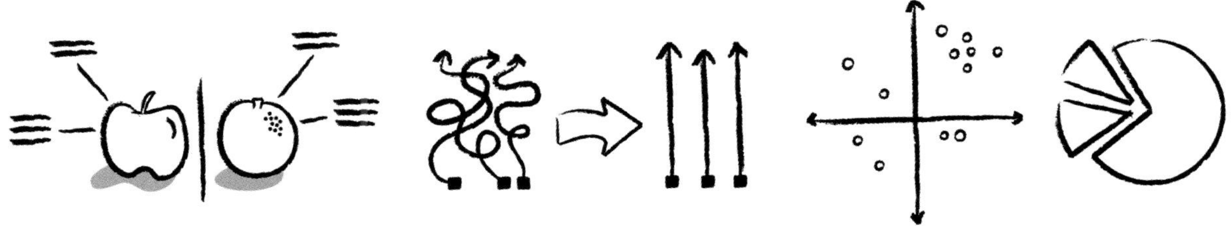

VISUALIZE YOUR OPPORTUNITY

Create the story of discovering your opportunity using the Opportunity Report.

OVERVIEW

Sharing your progress and getting feedback is an important step in your opportunity journey. Using the Opportunity Report you can more effectively tell your story, get input, and evolve your thinking before moving to the next phase of crafting your strategy.

TIME NEEDED

3–4 hours, or more

MATERIALS

Markers, sticky notes, whiteboard, Growth Initiative Brief, and ALL prior activity material and output

STEPS

1. Bring all the materials from the Opportunity phase: Growth Initiative Brief, observation material, results from insights workshop, resource and ecosystem insights, user journey map, and Breakthrough Question. Print out the Opportunity Report or download it from our site (www.theartofopportunity.net).
2. Work through each phase to tell the story of discovering your opportunity. For each step, you will want to communicate four elements: method, findings, insights, and decisions made. Work your way through the process of how you arrived at this opportunity.
   - What did you do to gain an understanding of your customer and noncustomer?
   - How did you identifying barriers to consumption and hurdles to satisfaction?
   - Where did you look for your opportunity? How did you make observations and what did you do to make sense of them?
3. What was your process to understand your firm, its assets, resources, and capabilities, and what did you find?
4. Describe how all of your discoveries led to the framing of your Growth Initiative Brief. Finish the Opportunity Report by entering your Breakthrough Question.
5. As a team, brainstorm how you can best tell your story. Use storytelling and visual frameworks within the elements of the Opportunity Report. See the book for ideas and suggestions on storytelling and visual frameworks. Don't feel constrained by the template. You don't have to "stay within the lines." What's important is clearly relating where you went on your opportunity journey and how you arrived at your Breakthrough Question.

Additional activity resources and templates can be found at www.theartofopportunity.net.

SOURCE: *The Art of Opportunity* authors

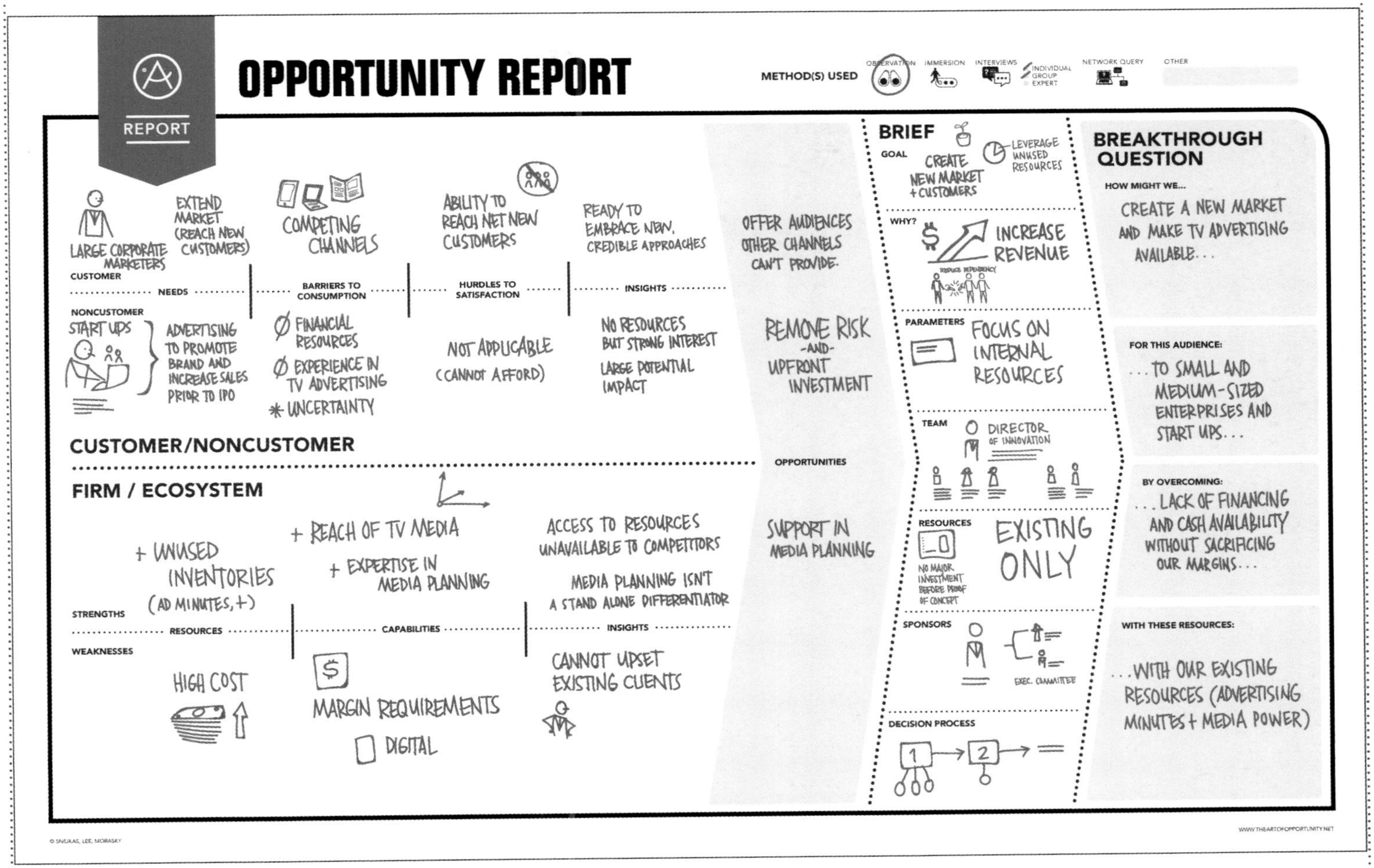

OPPORTUNITY REPORT

METHOD(S) USED

CUSTOMER/NONCUSTOMER

	NEEDS	BARRIERS TO CONSUMPTION	HURDLES TO SATISFACTION	INSIGHTS
CUSTOMER: LARGE CORPORATE MARKETERS	EXTEND MARKET (REACH NEW CUSTOMERS)	COMPETING CHANNELS	ABILITY TO REACH NET NEW CUSTOMERS	READY TO EMBRACE NEW, CREDIBLE APPROACHES
NONCUSTOMER: START UPS	ADVERTISING TO PROMOTE BRAND AND INCREASE SALES PRIOR TO IPO	Ø FINANCIAL RESOURCES Ø EXPERIENCE IN TV ADVERTISING * UNCERTAINTY	NOT APPLICABLE (CANNOT AFFORD)	NO RESOURCES BUT STRONG INTEREST LARGE POTENTIAL IMPACT

FIRM / ECOSYSTEM

	RESOURCES	CAPABILITIES	INSIGHTS
STRENGTHS	+ UNUSED INVENTORIES (AD MINUTES, +)	+ REACH OF TV MEDIA + EXPERTISE IN MEDIA PLANNING	ACCESS TO RESOURCES UNAVAILABLE TO COMPETITORS MEDIA PLANNING ISN'T A STAND ALONE DIFFERENTIATOR
WEAKNESSES	HIGH COST	MARGIN REQUIREMENTS DIGITAL	CANNOT UPSET EXISTING CLIENTS

OPPORTUNITIES

OFFER AUDIENCES OTHER CHANNELS CAN'T PROVIDE.

REMOVE RISK -AND- UPFRONT INVESTMENT

SUPPORT IN MEDIA PLANNING

BRIEF

GOAL: CREATE NEW MARKET + CUSTOMERS; LEVERAGE UNUSED RESOURCES

WHY? INCREASE REVENUE; REDUCE DEPENDENCY

PARAMETERS: FOCUS ON INTERNAL RESOURCES

TEAM: DIRECTOR OF INNOVATION

RESOURCES: EXISTING ONLY; NO MAJOR INVESTMENT BEFORE PROOF OF CONCEPT

SPONSORS: EXEC. COMMITTEE

DECISION PROCESS: 1 → 2 → =

BREAKTHROUGH QUESTION

HOW MIGHT WE... CREATE A NEW MARKET AND MAKE TV ADVERTISING AVAILABLE...

FOR THIS AUDIENCE: ...TO SMALL AND MEDIUM-SIZED ENTERPRISES AND START UPS...

BY OVERCOMING: ...LACK OF FINANCING AND CASH AVAILABILITY WITHOUT SACRIFICING OUR MARGINS...

WITH THESE RESOURCES: ...WITH OUR EXISTING RESOURCES (ADVERTISING MINUTES + MEDIA POWER)

3

CRAFT YOUR STRATEGY

WITHOUT CRAFTSMANSHIP, INSPIRATION IS A MERE REED SHAKEN IN THE WIND.

—**JOHANNES BRAHMS** *Composer*

Now that you have identified your opportunity, how are you going to seize it? As we've discovered in the previous chapter, the world has no shortage of great ideas, and opportunities abound. But only those willing to put in the time to master the art of opportunity—Steve Jobs, the Wright Brothers, and others—dedicate themselves to the unglamorous reality of hammering out how they'll deliver on their great idea or opportunity. Why? Quite simply, it's hard work.

This chapter is dedicated to helping you seize your growth opportunity. We will identify the elements of your growth strategy needed to tap into the dormant growth potential of your least satisfied customers and unexplored noncustomers. We will help you be the architect of a successful new business. We'll show you how to craft your strategy.

THE ELEMENTS OF YOUR STRATEGY

WHILE TRADITIONAL STRATEGY would have you focus on products and services, strategic innovation means you will carefully design your strategy as the elegant combination of the following three parts:

1. **Offering:** The mix of products, services, and the customer experience.
2. **Business model:** The way you operate and the activities necessary to do business.
3. **Revenue model:** Where the money will come from, how you set prices, and how payment is done.

Although the three parts of your growth strategy are presented in a sequential fashion in this chapter, innovation can come from each of them. Maybe you have an idea for an innovative revenue model, and you start at that part, and then move to develop your offering and business model. Maybe you have an idea for a new offering and need to design the fitting business model. Or perhaps you have an interesting business model idea and need to develop the revenue model to go along with it.

In practice, you are likely to cycle back and forth between the three dimensions as each component informs the other, and you will apply recursive review and evaluation to hone your strategy. At the end of the day, you need to make sure all three parts are integrated and support each other to create a compelling strategy.

Come in We're

OPEN

DESIGN YOUR OFFERING

YOUR OFFERING IS THE BLEND of your products, services, and customer experiences. Taken together, these three components of your offering create value for your customers by meeting their needs in a satisfactory manner.

PRODUCTS

Think of a product as a tangible or intangible good that you sell that can usually, or to some degree, be directly experienced—seen, touched, smelled, or tasted, as well as tested. Selling a product entails the transfer of ownership from your company to a buyer. With ownership, the buyer receives permanent access to the product.

Products require manufacturing and can usually be stored, either physically or digitally, before being sold and being used. Smartphones, computers, and tablets are examples of products, as are the software and apps you use on these devices.

SERVICES

A service is an activity you provide that cannot be stored or owned. Services are typically consumed at the point of sale and the moment they are provided. Representative examples of services are consulting, training, car maintenance, getting a haircut, going to the movies, renting a car, or flying from Paris to New York.

In recent years it has become popular to turn products into services. Software as a service (SaaS) is such an example, whereby you no longer buy ownership of the software but only gain access to it by paying a monthly or yearly subscription fee. Online music or video streaming services operate on the same concept. Turning a product into a service often requires a switch in revenue models, as we will illustrate later.

CUSTOMER EXPERIENCES

Customer experiences go beyond merely selling products and services, and focus on your interactions with your customers during all points of contact with your company, product, or service. These interaction points include every step along the customer journey: how customers become aware of your company, product, or service, as well as evaluation of options, purchasing, delivery, usage, purchasing supplements, maintenance, and disposal as outlined in Chapter 2.

As said earlier, designing an innovative offering can start with any of these components of your offering. Depending on your growth brief objectives, you may want to enhance or change your current product, introduce new products and/or services, or instead redesign your customers' experience, or maybe address all three components at once to create an entirely new offering.

INSPIRATION

HILTI ENHANCING THE CUSTOMER EXPERIENCE

Hilti, a global leader in solutions for professional customers in the construction industry, went from developing, manufacturing, and selling products to offering complementary services and a comprehensive customer experience to achieve new growth and competitive advantage.

THE PRODUCTS

Hilti is famous for the quality and performance of an innovative range of products for the building trade. The core business focuses on designing and manufacturing tools and equipment for multiple construction needs such as measuring, drilling, demolition, cutting and grinding, fastening, and installation.

THE SERVICES

In the 1990s, Hilti started to develop services to complement its product offering. Today these services range from providing design software and engineering solutions to product and application consulting and training. But the company is best known for its lifetime services and fleet management services.

Hilti offers three levels of warranty, starting with the basic warranty (during which tools are repaired at no cost) that comes with any purchase. As additional paid services, Hilti offers the lifetime repair cost cap, meaning customers pay a fixed percentage of the tool's list price for repairs, and the lifetime manufacturer's warranty, which provides the buyer with free repairs "due to defects in materials or workmanship . . . as long as replacement parts remain available over the entire life of the product."

With fleet management, Hilti gives customers the option to no longer buy tools. Instead, for a fixed monthly fee customers can rent the tools they need. In addition to covering all tool, service, and repair costs, Hilti manages the customers' tool inventory, making sure that they have tools when and where they need them. Tools in the fleet are replaced regularly, providing customers with access to the newest tools without having to invest. As a result, customers benefit from a reduced cash expense, less administrative effort for purchasing tools, and no administration of existing tool inventories. Additionally, Hilti found fleet management customers to be more loyal to the brand and to use more products, which ultimately increases their spend with Hilti.

THE CUSTOMER EXPERIENCE

Controlling the customer experience has always been important to Hilti. It has never distributed tools through third parties, relying instead on its own direct sales force. Even if it sells tools at local hardware stores, Hilti relies on a shop-within-shop experience, where Hilti's technical sales advisers are in charge of the customer experience. Over the years, Hilti has introduced a number of activities and services that enhance the customer experience and address the most important steps in the customer journey.

Besides the fleet management services outlined, which can dramatically enhance a customer's experience, Hilti also offers a range of logistics and financial administration services. For example, thanks to Hilti's in-house logistics services, with a Hilti online account, ordering new tools can be completed online, payment methods can be chosen, invoicing happens electronically, and tools can be delivered within one or two days. Hilti's eProcurement solutions allow electronic data exchange and online eCatalogs to be integrated into a customer's enterprise resource planning system, automating much of the ordering, billing, dispatch, and payment processes.

With its products, services, and solutions across the entire value chain, Hilti addresses the needs of various customer groups and stakeholders involved throughout the buying and usage process.

OFFERING SPARKS

10 WAYS TO DESIGN A NEW OFFERING

Here is a set of "Offering Sparks" to prompt your imagination, so you can explore options for designing your offering. These tactics are by no means exhaustive, and the hope is that you and your team will push your offering boundaries to find an optimal solution. Some of the tactics provide contradictory approaches—for example, "Spark 1: Enhance your existing offering" and "Spark 2: Reduce your existing offering."

The key here is not to try to apply all of them at once, but to try them out and see what your offering would look like if you applied only one tactic or maybe a creative combination of multiple tactics. For some companies and industries, enhancing the offering can be the way to go, whereas for others reducing it is more promising. While applying a tactic, always think about your customers and their needs! And to help visualize the offering options, enjoy our tasty analogies!

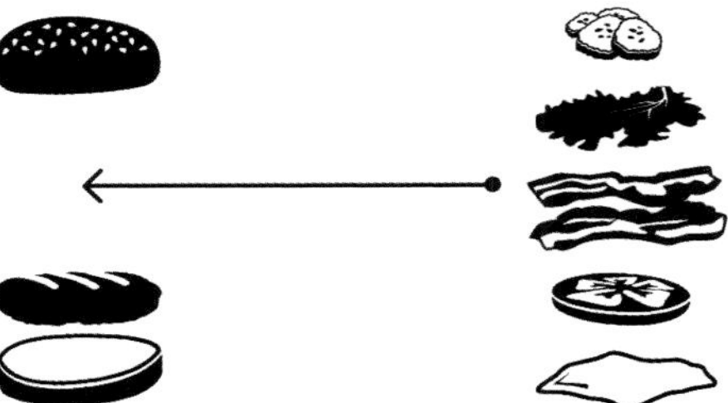

ENHANCE

Add extras to your burger.

1 ENHANCE YOUR EXISTING OFFERING

The first and probably most popular strategy to achieve growth is enhancing your current offering by upgrading your products, services, and customer experience with features your customers and noncustomers value. Let's say you run a railway company. A reason why noncustomers do not use your services could be the lack of comfort. Upgrading your trains with business-class or first-class services might convince noncustomers to switch to your offering. Product innovations and upgrades like these are typical examples of this strategy.

Enhancing your current offering could also mean enlarging the number of products and services you sell. Applying a long tail strategy to your offering would mean selling a higher number of products, but at a lower number of each product.

Questions to Consider:

- Which new features could be offered to customers and non-customers?
- Which features would make our offering more interesting to them?
- What products or services are customers and noncustomers buying instead of our offering? (Consider competitors, but also different industries.) Why?
- Which of these features could we integrate into our products, services, and customer experience?
- Which other products and services could we offer through our channels to apply a long tail strategy?

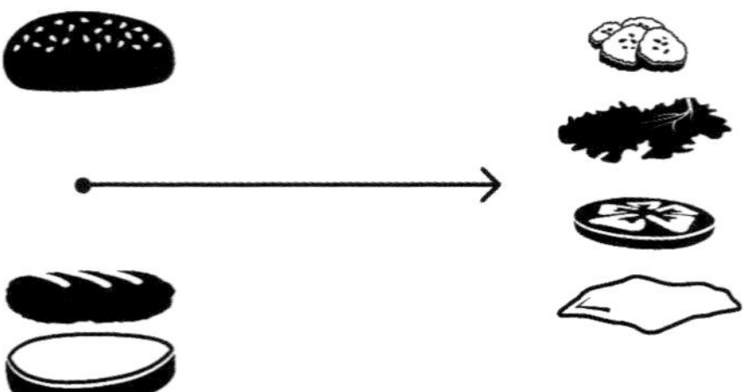

REDUCE

Take away extras from your burger.

2 REDUCE YOUR EXISTING OFFERING

Instead of adding new features, a new offering can consist of only the most necessary of features, which can be offered at a lower cost. Low-cost airlines like Ryanair or easyJet and budget hotels like Motel 6 in the United States or Motel One in Europe are probably the most prominent examples of going no-frills.

This strategy works particularly well if you have identified that noncustomers or refusing customers do not buy because your offering is too expensive and overdelivers on customers' needs.

The original idea of Clayton Christensen's disruptive innovation theory consists of offering products that are performing less than existing ones, but are nonetheless offering enough for a large percentage of the customers. No-frills offerings can also enhance the ease of use and hence make the products more readily available to a larger market and other customer segments that previously refused to use them because they were too difficult to handle or too expensive.

Questions to Consider:

- Which features do customers really need?
- Which features are customers willing to purchase?
- Which features that make our offering expensive are not valued by customers?

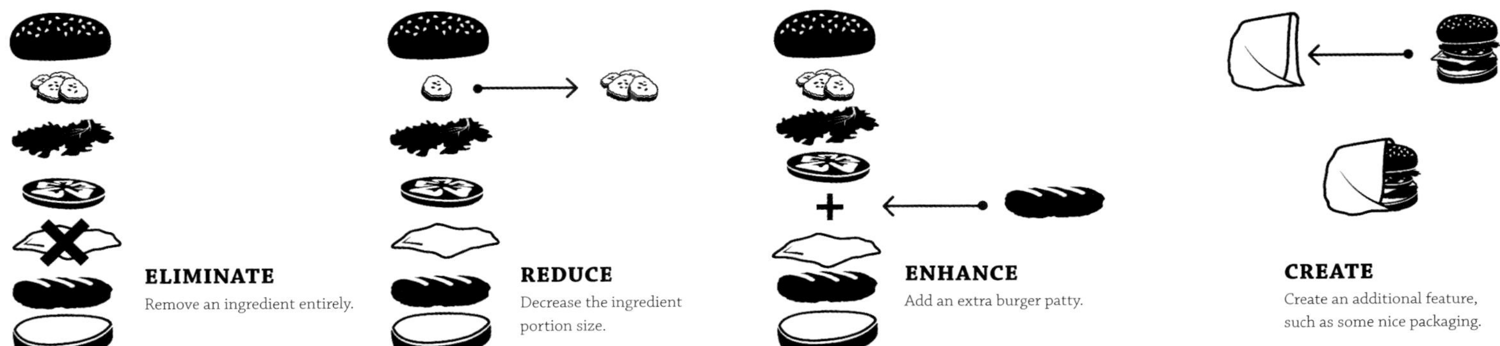

3 SELECTIVELY ELIMINATE, REDUCE, ENHANCE, AND CREATE (EREC) FEATURES AND YOUR OFFERING

Instead of either enhancing your offering or going no-frills, you can consider selectively eliminating, reducing, enhancing, and creating features as has been suggested by the Blue Ocean Strategy framework. Apple came back to success in the late 1990s by reducing its product range to only four products: two laptops (one for home users and one for professional users) and two desktops (again, one for home users and one for professional users). The EREC approach thereby creates additional value for customers, while reducing the cost for the company.

Questions to Consider:

- Which features that our industry has taken for granted can be eliminated or reduced (because they are costly and the majority of customers do not use them)?
- Which features and/or products, services, and experience elements could we introduce that would make our offering more interesting to customers and noncustomers alike?
- Which features could be reduced because they constrain consumption by making the product overly complicated to use?
- How can we simplify the customer experience?
- How can we simplify our offering and make it easier for our customers to choose?
- Which features would dramatically increase the value for our customers and noncustomers?

SWITCH YOUR APPEAL
Enhance your brand appeal with new packaging.

4 SWITCH YOUR APPEAL

Switching the appeal of your offering from functional to emotional or vice versa can make the same offer more compelling. Style and design can play an increasingly important role when it comes to products, services, and the customer experience. User interfaces need to be well designed to make websites, apps, and software products interesting. A cool, trendy product design can offer strong product appeal.

Consider Apple. Unlike other computer manufacturers, Apple does not put a lot of emphasis on the technical specifications (e.g., processing speed or hard disk capacity) in its advertisements, but rather positions its products based on emotions. All these techniques emphasize the emotional appeal of products. Hilti's branded red tools are considered to be cool, and workers are proud to show off their Hilti hammers.

Questions to Consider:

- What is the dominant appeal of our industry?
- Does it focus more on technical specifications, customer service, design, or emotions?
- What would it look like to switch to the opposite?

COMPLEMENTARY
Products, such as drink and fries, add value.

5 OFFER COMPLEMENTARY PRODUCTS AND SERVICES

Offering product and service extensions, add-ons, and plug-ins is one of the most commonly pursued and highest-potential growth strategies. If your objective is to stay close to your core business, it is also a relatively safe and low-risk tactic. Examples include smartphones plus charger, headphones, and all sorts of adapters; fast-food restaurants offering fries with a burger; hotels teaming up with restaurants or spas; suggesting an extended warranty when selling a product; automotive companies offering financial services; or manufacturers offering installation and maintenance services with their products.

Questions to Consider:

- What complementary products and services would enhance the customer experience along the customer journey?
- What products can we offer along with our services?
- What services can we offer along with our products?
- Which complementary services would make our customers' lives easier and the buying experience a lot more satisfying?
- Which other products do our customers typically buy along with ours, but have to get from somebody else?

A NEW EXPERIENCE
Customize your meal to your specification.

6 OFFER SOLUTIONS AND EXPERIENCES

Offering complementary products and services can be further enhanced by offering your customers a holistic solution and experience. Consider the example of Emirates Airlines, which offers business-class and first-class passengers a pick-up and drop-off service, taking them by chauffeur from their home to the airport and on arrival from the airport to their final destination. The need is to arrive at your final destination and not just at the airport, after all.

Mercedes-Benz's Car2Go service goes beyond renting a car as a service to providing a mobility solution. Its Mercedes Me, or similar offerings, like BMW's DriveNow service, take it even a step further, not only offering you different types of cars for different purposes, but also teaming up with public transportation companies to offer a holistic transportation solution, and with car rental companies to provide access to cars in more cities.

Questions to Consider:

- Looking at our customers' total buying experience: What it is that they are trying to accomplish? What is the job to be done?
- What solutions can we offer to accompany the buying experience? What could a holistic solution along the customer journey look like?
- How can we engage our customers in co-creation experiences?
- Could our customers personalize or design products or parts of products themselves?
- Why do people buy our product—to own it, or because they want to solve a problem or satisfy a need?
- What could an integrated offering, making the customer's total experience more worthwhile, look like?

BUNDLE

Bundle an additional burger for a bigger deal.

7 OFFER BUNDLES

Offering bundles is a similar strategy to offering complementary products; however, with bundles you usually do not have a choice. The bundled products and services often lock in customers to your offering. Think about razors and blades. You cannot use one without the other. Computers come with operating systems; mobile phones come with contracts; combo meals come with burgers, fries, and a drink.

Obviously, you want to create bundles of products and services that customers often consume together anyhow or that they require to satisfy a certain need. All-inclusive vacations with flights, hotel, rental car, and full board and room at resorts are a good example.

Amazon Prime is a good example of a bundle: For an annual fee you get faster delivery, you can rent books, and you get video-on-demand access. Or think about media companies offering integrated TV, plus Internet, plus landline, and cell phone contracts. Such bundles increase your share of wallet, while enhancing the customer experience by reducing the hassle of dealing with multiple companies for these services.

Questions to Consider:

- Which of our products and services can we integrate in a meaningful way?
- Which of our products and services do customers usually buy together anyhow?
- Which products do customers buy from competitors, although we offer similar ones?
- What is the total solution customers want to buy?
- Which bundles can we offer that address separate needs customers typically have?

EXTEND

Extend your customers' experience by expanding your offering.

8 EXPAND THE USE OF YOUR ASSETS AND CAPABILITIES

Although it is recommended that you look at your customers first, ideas for new offerings can also come from evaluating your assets and asking yourself how you can leverage them to create a new offering and potentially a new customer segment. Airlines and hotels are good examples of how assets can be leveraged to cater to the needs of different audiences.

Airlines use one plane to offer economy, business, and first-class services. Hotels operate on the same logic with their different types of rooms. McDonald's used its stores very successfully in Europe to establish McCafé as an offering catering to a different customer group. Maybe you own a building with unused office space that you could start renting out as a coworking space. Perhaps an unused factory area could become an event location.

Questions to Consider:

- What are our most valuable assets?
- What other ways can our assets be used? (List everything you could do with them!)
- What do we use them for today?
- How can we use them to address the needs of noncustomers?
- What other ways could we use our assets, even completely unrelated to their current use?
- Which of our assets, capabilities, and core competencies are truly unique?
- Which of those are valued most by our current customers?
- What needs do our assets and resources fulfill?
- Which customers have a similar basic need to ours (like getting from Paris to New York), but want a different experience?

DIGITAL
Provide digital access to see menus and reviews.

9 DIGITAL TRANSFORMATION: TURN SERVICES INTO PRODUCTS AND/OR PRODUCTS INTO SERVICES

Digital transformation has a lot to do with turning services into digital products. Think about the recent boom in fitness apps, for example. Apps like FitStar, FitYoga, Bodyweight Training, or Freeletics turn the services offered by regular gyms into digital products. McKinsey Solutions is another example of how to turn consulting services into online products. PSFK, a New York–based firm, has been labeled "the future of consulting." What it did was turn expertise and how expertise is provided from a service into an online product. In a similar vein, online education sites like Coursera turn educational services into digital products.

Software as a service (SaaS) is probably the most prominent example of how products can be turned into services. The previously mentioned services like Car2Go or Hilti's fleet management are other examples. Online video and music streaming services like Amazon Prime, Netflix, Spotify, or Apple Music turn buying products into online services.

Questions to Consider:

- How can our expertise be digitalized?
- How can our services be digitalized as do-it-yourself services?
- How can we leverage technology to make our products and services accessible to a wider audience?
- Which of our products do customers want or need to own?
- Which specific services need to be accessed by the customer?

CUSTOMIZE
Customize your meal — even in advance!

10 DESIGN YOUR CUSTOMER EXPERIENCE

During your search for new growth opportunities, you have probably discovered gaps in the customer journey—that is, steps the customer needs to complete—that the current offering in your industry is not addressing or that are particularly cumbersome or unpleasant for the customer to complete. Designing the customer experience relates as much to single steps as to the whole customer journey.

Outfittery has introduced a service whereby complete outfits get sent to customers, who can try them on at home, keep the pieces they like, and send back those they do not like. A service like Uber makes the experience of hiring a cab much easier: no waiting in line for a cab, or on the street for a taxi to pass, no fiddling around with your wallet, and easy retrieval of receipts to claim expenses.

Your brand image also has an impact on the customer experience. Are you perceived as the technology and innovation leader or rather the low-cost, no-frills company? Is your brand sexy or rather dull?

Questions to Consider:

- What does our customer's journey look like?
- Which steps of the journey do we address with our products and services?
- How can we make each step easier and more convenient for the customer?
- How do we reach our customers? Through which channels do they engage in business with us?
- How aligned are our channels and processes with the needs of our customers?

OFFERING BRAINSTORM

Using your Growth Initiative Brief and set of Offering Sparks, work with your team to generate offering ideas.

OVERVIEW

This workshop leverages Offering Sparks and a team-based approach to create ideas for products, services, and customer experience for your offering.

TIME NEEDED
90–180 minutes

MATERIALS
Markers, sticky notes, whiteboard, Growth Initiative Brief, Offering Sparks

STEPS

1. Print out or download a set of Offering Sparks Cards. Give each person a copy of the Growth Brief. Then pass out sticky notes, voting dots, and markers to all participants. If you don't have cards, print a page for each Offering Spark and its corresponding questions.
2. Write your Breakthrough Question at the top of the whiteboard. If you have more than one, duplicate the areas on the whiteboard. Then use three large sticky notes (or create three separate areas on a whiteboard/wall well apart from each other) and label them: Products, Services, and Customer Experience.
3. Divide into three teams, and assign them to one of the Offering component areas. Using the Growth Initiative Brief as a guide, along with the Sparks for inspiration, have participants think of new ideas for (or variations on) products, services, and customer experience.
4. Write all ideas on sticky notes, one idea per note, and post them in one of the corresponding categories from Step 2.
5. When done, allow everyone to briefly describe his or her ideas and allow time for group discussion. Remind participants that this is not intended to be an exhaustive exercise; rather it's an opportunity to get the first ideas on paper. In subsequent steps, additional ideas will be generated.
6. Each team then rotates to a new category area and builds upon/evolves the posted ideas or creates entirely new ones, recording their ideas with one idea per sticky note. When done, each team rotates again and repeats this process. Once ideation is complete, the entire group meets to review each category's ideas and further refine and develop them.
7. Once all the votes are cast, identify the top three ideas for each Offering component. Then have a group discussion to decide if you want to alter or revise the list. Take photos of all whiteboards.

Additional activity resources and templates can be found at www.theartofopportunity.net.

SOURCE: *The Art of Opportunity* authors

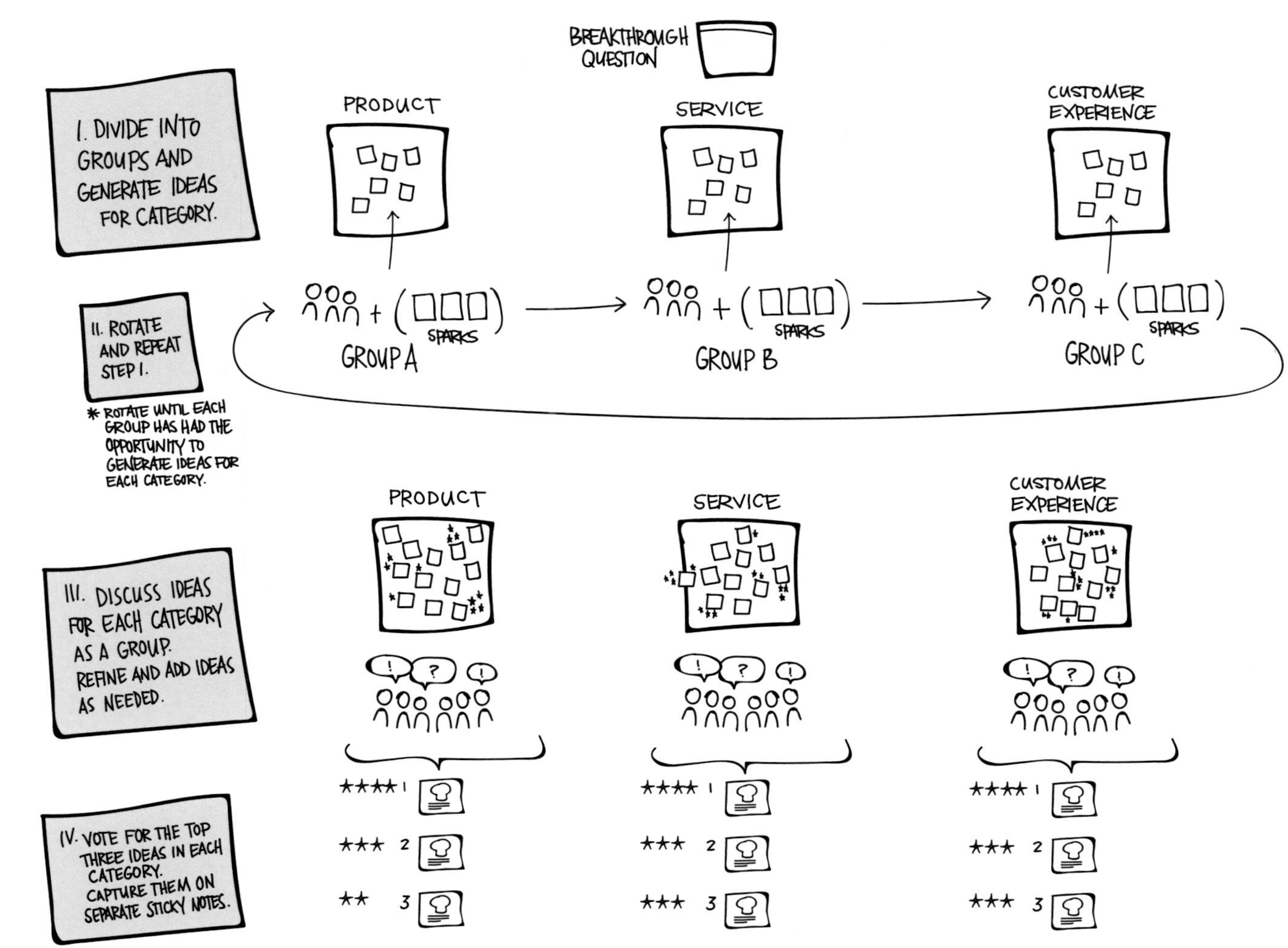
BREAKTHROUGH QUESTION
I. DIVIDE INTO GROUPS AND GENERATE IDEAS FOR CATEGORY.
PRODUCT
SERVICE
CUSTOMER EXPERIENCE
II. ROTATE AND REPEAT STEP I.
* ROTATE UNTIL EACH GROUP HAS HAD THE OPPORTUNITY TO GENERATE IDEAS FOR EACH CATEGORY.
SPARKS
GROUP A
GROUP B
GROUP C
III. DISCUSS IDEAS FOR EACH CATEGORY AS A GROUP. REFINE AND ADD IDEAS AS NEEDED.
IV. VOTE FOR THE TOP THREE IDEAS IN EACH CATEGORY. CAPTURE THEM ON SEPARATE STICKY NOTES.
★★★★ 1
★★★ 2
★★ 3
★★★ 3

DESIGN YOUR OFFERING

A combining activity to determine the components of your offering, as you craft the strategy of your new growth business.

OVERVIEW

From prior work, you have ideated and refined a set of options for each component of your offering. This team activity uses collaboration to see creative component combinations so you may choose the offering that best fits your Breakthrough Question.

TIME NEEDED

60–90 minutes

MATERIALS

Markers, sticky notes, whiteboard and Growth Initiative Brief

STEPS

1. Have a copy of your Growth Initiative Brief for each person, as well as the output from the Offering Brainstorm activity. Write your Breakthrough Question at the top of the whiteboard. If you have more than one, duplicate the areas on the whiteboard.
2. Then draw three columns on separate areas and create a label for Products, then one for Services and the last one for Customer Experience. Post up the three best (most innovative, completely unique) ideas from the Offering Brainstorm session in each corresponding offering component category.
3. Divide the group into teams. Give each participant sticky notes and markers. Now the fun begins. Knowing that your offering is the unique blend of products, services, and customer experience, have each group explore different combinations of the components in each category. And if a new inspired component emerges while combining, all the better. Try to visualize each component to increase understanding.
4. After exploring combination options, each team reviews the offering combination concepts with the group and together they discuss and come to an agreement on which offering combination bests satisfies the Breakthrough Question and the goals of the Growth Initiative Brief. Take photos of all whiteboards.

Additional activity resources and templates can be found at www.theartofopportunity.net.

SOURCE: *The Art of Opportunity* authors

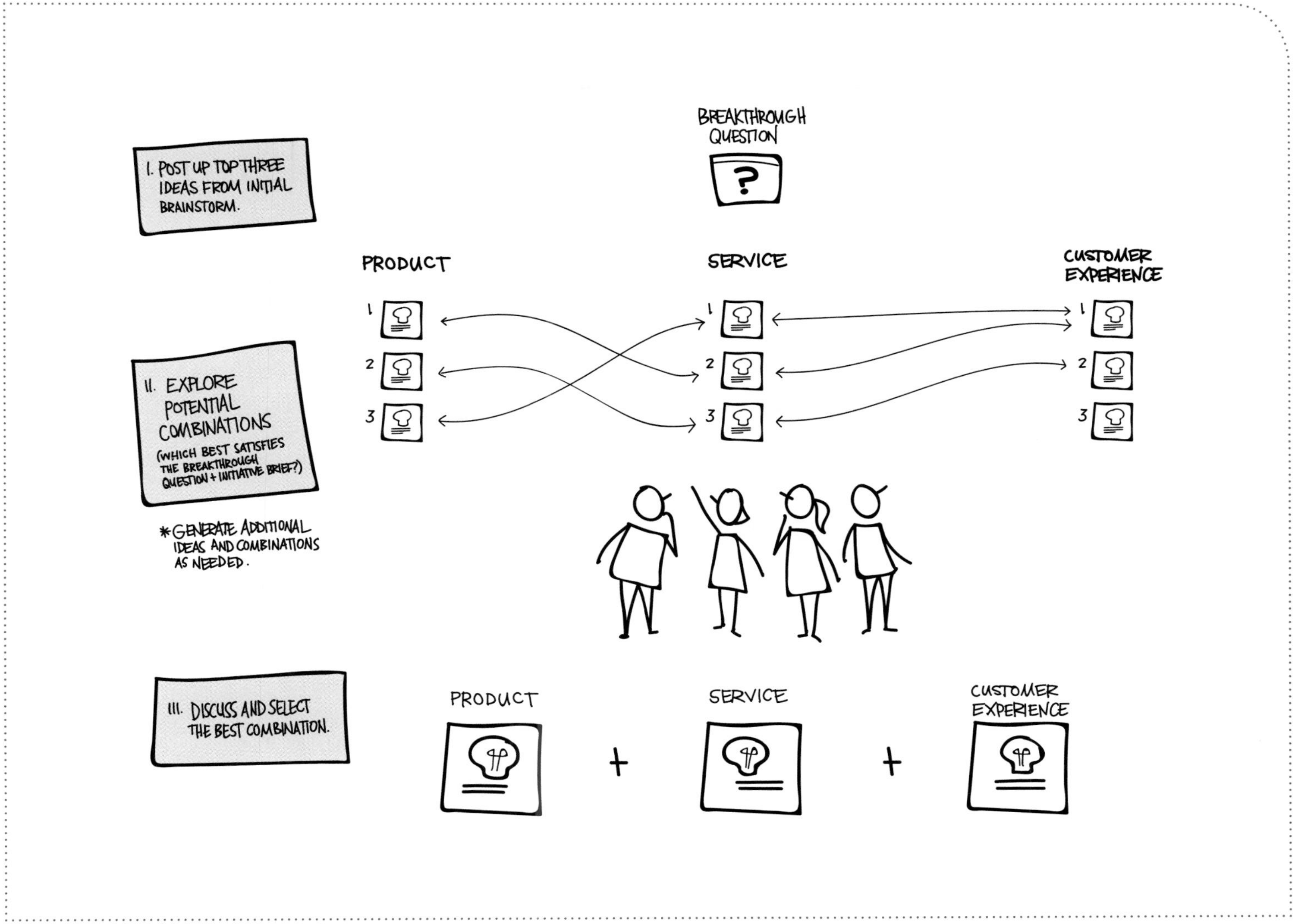

BREAKTHROUGH QUESTION

I. POST UP TOP THREE IDEAS FROM INITIAL BRAINSTORM.

PRODUCT

SERVICE

CUSTOMER EXPERIENCE

1
2
3

II. EXPLORE POTENTIAL COMBINATIONS (WHICH BEST SATISFIES THE BREAKTHROUGH QUESTION + INITIATIVE BRIEF?)

*GENERATE ADDITIONAL IDEAS AND COMBINATIONS AS NEEDED.

III. DISCUSS AND SELECT THE BEST COMBINATION.

PRODUCT + SERVICE + CUSTOMER EXPERIENCE

GET THE BUSINESS MODEL WRONG, AND THERE IS ALMOST NO CHANCE FOR SUCCESS.

—**DAVID TEECE** *Professor and Organizational Theorist*

SHAPE YOUR BUSINESS MODEL

IN RECENT YEARS THE CONCEPT of the business model has raised considerable interest. Starting with the dot-com boom, it slowly found its way into the mainstream, first in managerial practice, and later in academia. Despite all that has been written and said on business models, there is still confusion around this poorly understood concept. Consequently, executives do not have a shared understanding of the concept and are starting to lose interest.

The problem stems in particular from the multitude of definitions of what a business model is, and the differing views of the concept. Let us bring some clarity by illuminating four distinct perspectives on the business model.[1]

SHAPE YOUR BUSINESS MODEL

THE STATIC PERSPECTIVE

The static perspective focuses on answering the question: What is a business model? This perspective relies on definitions, components, and building blocks, and provides conceptual, textual, and graphical or visual descriptions and representations.

The many definitions of the static perspective can be distilled to the following primary components:

- Financials, revenues, profit, pricing, cost
- Resources, assets, capabilities, competencies
- Activities, processes
- Strategy, competitive advantage, differentiation, positioning
- Value proposition, benefits, solutions
- Network, partners, suppliers, ecosystem
- Customers, customer segments, target market
- Offering, products, services
- Governance, relationships, collaboration
- Organization

There are two variants of definitions: the broad definitions, using a large number of components to define the business model, often including customers, the offering, the revenue model, and so on; and the narrow definitions, defining the business model as an activity system and focusing on the activities to be performed, who performs them, how they are linked and sequenced, and the assets and skills necessary to perform them.

THE DYNAMIC PERSPECTIVE

The static perspective focuses on describing business models, but the dynamic perspective takes a process view and is interested in how change and innovation of business models happen. Two approaches explain change and innovation processes: contextual and rational.

Proponents of the contextual approach believe that business model innovation happens if the required conditions, a favorable culture, the right organizational structure, and the right leadership styles are present. Proponents of the rational approach believe that management and organizations can engage in specific activities and processes to actively design new business models.

These activities usually follow these four steps:

1. Understanding customer needs and the business environment through analysis.
2. Developing the new business model.
3. Evaluating business model ideas, often through experimentation.
4. Implementing and scaling up the new business model.

THE STRATEGIC PERSPECTIVE

The strategic perspective is interested in how business model innovation and change can create value for key stakeholders: customers, the firm itself, and its ecosystem. It sees the business model as a distinct strategic choice that has to be aligned with product and market decisions. Whereas some believe that small changes are sufficient, others believe that radical innovation is necessary to create new and unique value. Different business models can open up new strategic opportunities for growth and venturing into new business areas. Consider Amazon, whose online business model enabled it to venture from books into other product areas and completely new business areas, opportunities that the typical bookstore business model does not offer.

THE OPERATIONAL PERSPECTIVE

The operational perspective focuses on how to operate, manage, and control business models effectively and efficiently once they have been implemented. Authors differ as to whether the goal of this management is the continuous improvement to increase efficiency or the management for growth.

OUR DEFINITION OF A BUSINESS MODEL

We apply the narrow definition of a business model as a holistic system of activities needed[2] to create and deliver your offering, resources necessary to perform these activities, the processes and the sequence of how these activities are being performed, and the organizational units and external partners who perform these activities.

SHAPE YOUR BUSINESS MODEL

BUSINESS MODEL COMPONENTS

ACTIVITIES

Which activities need to be performed as part of your business model, and how will they be performed? To design your business model, it is important to consider and describe all the activities needed to operate your business and to create and deliver your products, services, and customer experiences, no matter whether you will directly perform these activities or will outsource key parts. Selecting which activities need to be performed requires you to think about a number of aspects. Which activities are required to deliver an experience that best satisfies the customers' needs?

Let's look at Amazon. Getting products from its warehouses to the customers is a vital part of the business model, without which Amazon's model would not work. Yet Amazon outsources delivery to partners like FedEx, UPS, or DHL. Or think about Procter & Gamble's open innovation initiative. Researching and developing new products is key to Procter & Gamble's business model, and the company's Connect + Develop initiative is focused on sourcing innovation ideas outside of the company through open innovation, instead of doing everything in-house.

You might also think about the form a certain activity takes. Warehousing at Amazon looks different from warehousing at Barnes & Noble. For some activities, it is necessary to specify in greater detail what they will look like.

ACTIVITIES TO CONSIDER IN DESIGNING YOUR BUSINESS MODEL

CUSTOMER MANAGEMENT	Customer selection/targeting, acquisition, development, retention and growth, marketing, sales, and public relations.
INNOVATION/DEVELOPMENT	Research and development; designing products, services, and customer experiences; programming
REGULATORY AND SOCIAL	Environment, health and safety, governmental affairs, community relations
INFRASTRUCTURE AND ORGANIZATIONAL	Talent selection, recruiting, and management (HR); IT infrastructure management and maintenance; operating servers, websites, databases, platforms, apps, etc.; data collection and analysis
FINANCE AND ADMINISTRATION	Finance, accounting, administration[3]

SHAPE YOUR BUSINESS MODEL BUSINESS MODEL COMPONENTS

RESOURCES

Which resources, assets, skills, and capabilities are required to perform the activities necessary for your business model? Irrespective of who provides or owns the assets and who performs the activities, how the chosen activities are performed will require you to think about the assets and skills necessary to perform those activities. Consider a typical restaurant or retail franchise, for example. It needs access to prime real estate locations to be able to perform its sales activities well. The real estate can either be bought or be rented. In any case, access to real estate is vital.

Again, it is not important for your organization to own the required resources or skills, but you need to think about which ones are essential to your business model. Think about online platforms like Airbnb or eBay. Having rooms and products on the platform is key to the business model, yet neither Airbnb nor eBay owns these resources. In fact, Airbnb is the largest accommodation provider on the planet, without owning any rental property itself.

Typical resources to think about:

- People and talent
- Skills and capabilities
- Knowledge, data, and information
- Technology and patents
- Brand and image
- Equipment, infrastructure, and production facilities
- Channels
- Exclusive or privileged access to partnerships, alliances, and specific customer segments
- Financial resources, assets, cash flow, capital, and access to financial markets
- Location

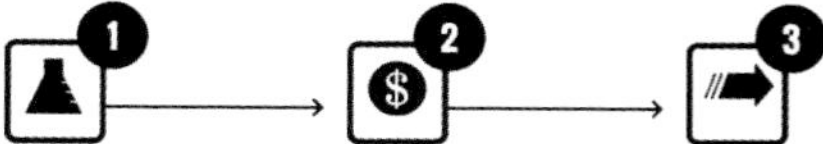

SEQUENCE

The sequence of your business model's activities determines how you link activities and in which order you perform them. An innovative business model can emerge from simply switching the order of activities within your company's or industry's value chain. A traditional sequence of activities in a manufacturing business would, for example, look like this: sourcing of raw materials, warehousing of raw materials, manufacturing, warehousing of finished goods, distribution to retail space, and sales.

Dell was the first computer company to turn this sequence upside down. First, customers ordered and paid for their computers online. Only then would Dell source the required raw materials, manufacture the computers, and ship them directly to the customers. By shipping goods directly to the customers, a key link—the retailer—was eliminated, and carrying material inventories became unnecessary. The just-in-time sourcing of raw materials eliminated the need to carry large inventories.

Or ponder the global automotive business models. In the United States, cars are manufactured to stock and sold directly at the dealer, whereas in Europe cars are mostly built to order. These are the same activities, yet are performed in a different sequence to fulfill the specific customer needs of a certain region.

Typical processes in a sequence to think about:

- Sourcing, procurement, and purchasing
- Inbound and outbound logistics and delivery
- Inventory, warehousing, and stock management
- Operations, manufacturing, assembly, and packaging
- Quality control
- Innovation, research and development, and product/technology development
- Marketing and sales, customer relationship management, and market research
- Sales, order processing, invoicing, and payment
- Human resources, hiring, and staffing
- After-sales services, support, and warranties
- IT processes
- Legal, financial, and administrative functions
- Facility management and maintenance

SHAPE YOUR BUSINESS MODEL BUSINESS MODEL COMPONENTS

ROLES

The final element of your business model is about who performs which activities and who provides the assets and capabilities necessary to do so. The key questions are: Which activities are vital for your business model? Which resources does your organization possess to fulfill these activities? Who performs the required activites and who provides the necessary resources?

If it would be too costly to build up the required resources and a specialized partner can better perform the activity, it might be better to outsource it. Let's look at Amazon again. Existing logistics companies can better perform delivery of products, and it would have been very costly for Amazon to build its own distribution network.

When it comes to working with partners, you will want to think about your relationship with these partners and how you orchestrate the collaboration. Do you want to have one exclusive partner and develop a close relationship, or do you want to outsource activities to many partners?

BUSINESS MODEL IN PRACTICE

To illustrate the business model as an activity system framework and how different business model designs impact one's business, look at the example of online or offline commerce. If you decide to be in the retail business, you can choose from at least two different business models:

1. Sell through a bricks-and-mortar store.
2. Sell online through the Internet.

The two options require different activities to be performed, and different resources, assets, and skills to perform them. If you decide to sell through a bricks-and-mortar store, you will need resources such as a store location, with merchandising displays, inventory, and salespeople working in the store. To sell online requires servers, possibly a warehouse, and people with different skill sets (e.g., programmers, warehouse workers, and customer service), plus a way to distribute your goods to your customers.

In the case of Amazon, the firm originally sold books (a well-established product) in the United States (a well-established market). Amazon changed the way books are sold, thereby creating a business model innovation, using the Internet to distribute its products versus the established bookstore retail business model, which made it easier and more convenient for customers to buy books and have them delivered to their homes. This business model required Amazon to operate a different activity system compared to the traditional bookstore.

BOOK SALES: RETAIL AND INTERNET BUSINESS MODELS

	TRADITIONAL BOOK STORE RETAIL BUSINESS MODEL	INTERNET RETAIL MODEL
KEY ACTIVITIES	· Establishing local bookstores (location search, rental, merchandising) · Local inventory management · Distribution to local stores · In-store displays · In-store services · In-store sales	· Central warehousing · Logistics · Website and server management · Online marketing
REQUIRED RESOURCES	· Bookstores · Local staff · Marketing	· IT skills · IT infrastructure · Central warehouse locations · Logistics in collaboration with outsourcing partners
PROCESSES SEQUENCE & LINKS OF ACTIVITIES	Ordering books, distribution to local store, local sales	Inventory management, direct distribution to customer, new linking as the bookstore was eliminated
STRUCTURE	All managed in-house	Distribution outsourced

INSPIRATION

EDEN MCCALLUM THE CONSULTING BUSINESS MODEL REDEFINED

Liann Eden and Dena McCallum discovered an opportunity to build a new type of consulting company based on two trends they observed. First, clients were getting more sophisticated in buying consulting. Many clients had been consultants themselves and did not need large consulting teams, but instead wanted expertise from senior professionals in very specific areas. Second, many of their consulting colleagues wanted to work in consulting, but no longer within the boundaries of one of the big consulting firms.

Many consultants just wanted to do consulting work and not be burdened with the administrative and sales tasks that are expected of them in the big consulting firms. Some also wanted to work part-time, something that is difficult to achieve in the traditional professional services firms. Eden and McCallum took advantage of these two insights to create a business model combining traditional strategy consulting and freelancing.

HOW DOES THE BUSINESS MODEL WORK?

The activities carried out in the Eden McCallum business model are not that different from those of the traditional consulting business model. Customers need to be acquired, and projects must be sold, staffed, and carried out. Yet the way these activities are organized at Eden McCallum is fundamentally different.

For starters, Eden McCallum does not employ any consultants. Each and every one of its more than 500 consultants is independent. Not having consultants on the payroll provides all parties with benefits. Not having to pay the fixed cost of employed consultants, Eden McCallum is under no pressure to sell the next project and to staff all its consultants onto these projects. Consultants, in turn, get to choose the projects they want to work on, just like clients get to choose the consultants they would like to work with.

Partners at Eden McCallum do not manage the client projects, but instead leave this task to the consultants. Consultants are not pressured to sell new projects, as this is the job of the partners, and hence can focus on delivering successful projects.

The lower overhead cost translates into lower fees for the customers, while the consultants can still earn a considerable salary from their freelance work. They enjoy the flexibility to decide how much they would like to work. Nobody pressures them into accepting the next consulting project.

As the consultants are highly independent in how they work with the clients, selecting and to some extent managing consultants is an important activity within the Eden McCallum business model. After all, the network of consultants is one of Eden McCallum's most valuable assets. A special team is in charge of building the talent pool, taking care of consultants, and managing all consultant-related activities.

Meanwhile, partners can focus on business development and building strong client relationships.

INSPIRATION

KLINIK HIRSLANDEN REINVENTING THE HOSPITAL BUSINESS MODEL

Klinik Hirslanden in Zurich is one of the most exclusive private hospitals in Switzerland. Founded in 1932, it is part of the Hirslanden Private Hospital Group, which was formed in 1990 by the merger of several private hospitals, and has been part of the South African Medi-Clinic Corporation hospital group since 2007.

Hospitals throughout the world typically operate on either of two business models: the chief physician system, in which doctors are employed by the hospital and report to a chief, or the private practitioner system, wherein doctors are independent and not necessarily strongly tied to a particular hospital.

HOW DOES THE BUSINESS MODEL WORK?

Over the course of five years, Klinik Hirslanden gradually changed its old business model of an infrastructure provider to system provider. This innovative and unique business model combined the chief physician and private practitioner systems, allowing Hirslanden to move from treating private patients only to providing medical care to publicly and privately insured patients alike, while being allowed to provide highly specialized medicine under the Swiss legal health care system. This transition required a shift from only providing infrastructure to doctors, to offering a full range of services to private practitioners and patients alike.

To begin design of its new business model, Hirslanden asked why and how patients choose a particular hospital. It found that it is often the family doctor or the specialist surgeon who refers patients to a particular hospital. In addition to patients, doctors were also found to be customers for the hospital. To offer these two customer groups the best service while providing patients with the safest service possible, Hirslanden employed generalist doctors and provided required core primary care services, like a 24/7 emergency unit, patient dispatch, general internal medicine, general surgery, radiology, anesthesiology, and intensive medicine, therapy, and nursing.

To govern all these activities, orchestrate the processes, and work between employed generalist doctors and independent private practitioners, Hirslanden created a new organizational structure, regrouping all medical-related services into the so-called Medical System. The Medical System is led by a doctor, who, unlike in the traditional chief physician system, is relieved of his medical duties, focusing exclusively on the management of the Medical System. The Medical System provides services to patients and the private practitioners alike.

Private practitioners are supported in setting up their practices, often in facilities owned by the hospital, and are organized within the so-called umbrellas, interdisciplinary centers of competence focusing on a medical need perspective, instead of medical disciplines.

To offer its private patients a unique customer experience, Hirslanden introduced a hospitality unit, regrouping all nonmedical patient services, comparable to those of a five-star hotel.

KLINIK HIRSLANDEN BUSINESS MODEL TRANSFORMATION

	KEY ACTIVITIES	PROCESSES	STRUCTURE
FORMER	• Focus on treating private patients • Provide infrastructure to private practitioners	—	• Chief physician system and organizational structure • Organize by area of expertise • Mix of employed generalist and specialist doctors, plus private practitioners
NEW	• Focus on activities for patient well-being and satisfaction • Treat private patients, plus patients under the legal health care system • Provide solutions to private practitioner surgeons • Orchestrate work within the newly created Medical System • Hospitality and medical program management • Support private practitioners in their strategy, organization, and marketing	• New patient process • New processes between departments • Standardize processes and quality standards to be followed	• Combine the chief physician system and the private practitioner system • New org design based on the Medical System, grouping general medical services • New hospitality unit regroups nonmedical patient services • Group medical centers and private practitioners around clinical needs • Employ generalist doctors • Specialists act as private practitioners

BUSINESS MODEL SPARKS

8 WAYS TO DESIGN A NEW BUSINESS MODEL

Setting a new business model or redesigning your current model requires you to contemplate your activities, resources, sequence, and roles for offering a complete, well-integrated model. Next is a set of "Business Model Sparks" to inspire you to find innovative options for designing your business model. When designing your new business model, keep the insights you gained about your customers and noncustomers, their needs and experience, as well as your company and its ecosystem in mind.

We suggest you use the different sparks one by one and try to design various options of your business model. Don't hesitate to go to extremes. Extremes often hold some elements that make a lot of sense. Once you have developed multiple alternatives, you might also want to try combinations. Play around and see what ideas come up.

ASSESS
Assess your assets and make your own hamburger buns.

1 ASSESS YOUR EXISTING BUSINESS MODEL

A good starting point to design a new business model is to evaluate your current one. Many companies do not understand their current business model well. Take the value of your current business model that you created, its underlying key activities, assets, and processes, and try to understand what is particular about it. Is there anything that makes it special? Also, look at different business models in your industry. Inspiration can come from combining the best of different models, as shown with Eden McCallum and Klinik Hirslanden. Look at the opportunities you identified, and think about what could be different in your business model to capture the opportunity and better fulfill customers' and noncustomers' needs.

Questions to consider:

- What are our most valuable assets?
- Which activities in the value chain do we perform best?
- Which activities in the value chain are key to being successful in our industry?
- Which are the dominant business models in our industry?
- What are the advantages and disadvantages of each model?
- How well do the activities we perform fit together?
- How well do the activities reinforce each other?
- How well do they fit with what our customers expect?

NEW ACTIVITIES
Add home delivery or take out.

2 CREATE NEW ACTIVITIES

New business models and growth can emerge from performing new activities. For Amazon, introducing the possibility to sell used books as a new activity brings about 20 percent of book sales revenues these days. Or think about the possibility to send items back to Amazon at no cost for the user, a new activity that has dramatically enhanced the customer experience and removed barriers to online shopping. Adding new activities can also mean integrating steps in the value chain, which have traditionally been performed by suppliers. Integration thereby refers to adding upstream and downstream activities in the value chain.

Questions to consider:

- Which activities do we need to add to fulfill our customers' needs?
- Which of these do we need to perform, and which are part of our core capabilities?
- Which activities do not provide any value for the customer?
- Which new activities in the value chain would offer an enhanced customer experience?
- Can we eliminate unnecessary activities in the value chain?
- Which activities are obsolete and can be eliminated?
- Which activities could we outsource, potentially to customers?

UNBUNDLE

Unbundle your supply chain by ordering pre-made burgers.

3 UNBUNDLE YOUR BUSINESS MODEL

While performing new activities can mean adding activities, new business models can also emerge from unbundling existing business models. In the traditional business model of telecommunications providers, there are three main activities: (1) developing the network, (2) marketing, sales, and customer contract management, and (3) developing content offerings, for example, music services.

The business model of Bharti Airtel, the leading telecom provider in India, is based on unbundling the traditional mobile telecom business model. It created a new business model by focusing on its core capabilities, and outsourcing building its network, as well as content production, to partners. Through outsourcing, Bharti Airtel focused its resources on core differentiating activities. In only paying network providers for used capacity, the business model allowed Bharti Airtel to turn fixed costs into variable costs, giving it financial advantages, without having to invest in expensive networks.

Rethinking your activities might simply entail eliminating activities. Consider IKEA. It has built a business model around eliminating the delivery and assembly of products and outsourcing these activities to customers.

Questions to consider:

- What are our core capabilities?
- Which activities could be better performed by somebody else?
- How could unbundling our business model help us to focus our resources (human and financial) on our core capabilities?
- Which parts of our business create real value for the customers and differentiation for us?

COLLABORATE
Collaborate with local producers by adding an organic offering.

4 COLLABORATE WITH SUPPLIERS, PARTNERS, YOUR NETWORK, AND YOUR ECOSYSTEM

As your business model is likely to require activities that are not your core capabilities, you will need to collaborate with your ecosystem of suppliers and partners. Some businesses wouldn't be possible without partners. Successful business model innovators don't see partners only as suppliers, but try to establish real value-adding relationships with their ecosystem stakeholders.

Study some of the examples that have been illustrated so far: Amazon needs to partner with logistics companies to deliver goods, Bharti Airtel couldn't do business without the network providers, and Eden McCallum's and Hirslanden's business models wouldn't be possible without the strong relationships to their independent partners. Tactics like crowdsourcing rely on collaborating with large communities. Platform-based business models like Uber, eBay, Airbnb, or Eden McCallum rely on a community on both the supply and the demand sides, without which their businesses wouldn't work.

Questions to consider:

- Which activities do we perform best?
- How does our cost basis change when we use suppliers and partners instead of internal resources?
- Which activities are not that important for us to perform?
- Do we have unique access to partners?
- How well do we use the connection with our ecosystem stakeholders?
- How could we intensify our relationship with partners to create a unique business model?
- If we were to start anew, which activities would we perform and where would we collaborate with partners?

NEW SEQUENCES
Instead of paying at the end, pay before getting the burger. No service, but take out your meal.

5 ESTABLISH NEW AND DIFFERENT LINKS AND SEQUENCES BETWEEN ACTIVITIES

Changing the sequence of how activities are performed is another strategy to design new business models. As previously discussed, Dell changed the traditional manufacturing sequence to sell first, before manufacturing started. The fashion retailer Zara changed the traditional business model from the usual two seasons (spring/summer and autumn/winter) to a new cycle every month. Data on sales is thereby collected and fed back to manufacturing and logistics. So instead of stocking garments for a whole season in a shop, Zara delivers to its shops on a daily basis. The data is furthermore used to pick up trends influencing the design and manufacturing of the next garments.

The fast fashion cycles mean that Zara has new collections in its stores at least every month, in some cases even every week. Hence, customers return more often to see the new collections. Whereas the usual apparel producer's designs-to-sales cycle lasts about 1.5 years, the cycle for Zara is about 20 days. Process automation and on-demand production strategies are other ways to create new links between activities and hence new business models.

Questions to consider:

- Could we redesign core processes differently to improve efficiency and effectiveness?
- How can we use data to change our processes?
- What would our processes look like if we turned them upside down?
- How can we integrate our supply chain?

NEW USES
Rent out the restaurant during quiet afternoons for company workshops.

6 USE YOUR ASSETS AND RESOURCES IN NEW WAYS

New business model ideas can emerge from thinking about how to leverage existing assets in new ways. Think about airlines or hotels. The same asset, planes or hotel property, can be used to address multiple customer segments, with economy, business, and first classes, or single rooms, twin bed rooms, junior suites, and suites. Bharti Airtel and T-Mobile use their customer bases as channels to distribute additional products and services, like music streaming, for example, whereby German T-Mobile customers have access to Spotify.

Questions to consider:

- How can we use our assets to create new businesses?
- What are our core assets?
- What are the advantages of having these assets?
- What are we using them for at the moment?
- What other ways could we use our assets and resources?
- Which of our assets are truly unique and cannot be imitated nor substituted easily by others?

RETHINK
Rethink your channel by franchising your restaurant.

7 RETHINK YOUR CHANNELS

Many business model innovations arise from changing the way companies interact with their customers. Going direct, going indirect, establishing pop-up stores, franchising, and licensing are all strategies to consider. Dell and Amazon went direct by eliminating the distributors. Hilti does its own distribution to be able to control the customer experience, while competitors like Black & Decker or Bosch distribute their products through intermediaries. Airlines and hotel chains make pricing and their offerings more and more attractive on their own websites, and less interesting when bought through third parties.

Questions to consider:

- Where, when, and under which circumstances do our customers need/buy/use our products and/or services?
- How easy or difficult is it for our customers to find and buy our product?
- How easy is it to do business and interact with our company?
- How aligned are our channels and processes with the needs of our customers?
- How would we improve the customers' total buying cycle experience?
- How would we reach noncustomers?
- What new distribution channels and innovative points of presence could we develop?

DIGITAL TRANSFORMATION

Use digital transformation by offering online customized ordering.

8 DIGITAL TRANSFORMATION: TAKE ACTIVITIES ONLINE

Digital transformation of business models means taking activities online and enabling activities that were not possible before. Zara's supply chain management, for example, is made possible by technology and the integration of sales, design, and manufacturing processes. Businesses like Amazon, iTunes, Netflix, Spotify, Uber, Airbnb, eBay, Facebook, and so on wouldn't be possible without modern technology.

Live-Frischetheke enables customers to chat live with the butcher and see the actual product they will receive. They can ask questions about how much they need for a party of six, how to best prepare the meat, or how long it can be kept in the fridge. Through live video, the butcher can show the actual pieces the customers will receive, which avoids surprises and customers can trust they will have what they need for their party. Products are delivered in a vacuum packaging via express delivery, made possible only by partners like DHL or UPS having introduced fresh food delivery capabilities in recent years.

Questions to consider:

- Which activities could we take online?
- How could we make the customer experience more efficient and effective?
- Which activities would we need to take online to create a better customer experience?
- How can we leverage technology in a way to enable high tech and high touch?
- Do we rely on conventional channels?
- Do we offer (co-creation) experiences to customers?
- Is every step of the buying cycle addressed by the channels?
- How easy is it to do business with our company?

VISUALIZE YOUR CURRENT BUSINESS MODEL

Visualize the business model of your firm in order to create innovative variation.

OVERVIEW

Map your current business model using the individual components of your model: Activities, Resources, Sequences, and Roles. This activity can also be used to envision the model of a completely new business.

TIME NEEDED

60–90 minutes

MATERIALS

Markers, sticky notes, and whiteboard

STEPS

1. Give each participant sticky notes and markers. Have the team ideate the set of activities you currently perform to deliver your offering. For some new ideas, reference the table of activities shown in the book. Only write one activity per sticky note. Post them all on the whiteboard.
2. Grab the results of the Map Your Resources activity and post up the resources you own or have access to in order to fulfill the activities you outlined.
3. Now gather the set of roles you created in the Map Your Ecosystem activity. Assign roles to the activities and resources. For each activity, try to answer who performs it, and for each resource, who provides it. Roles can be internal groups or business units, and external stakeholders.
4. Next, create the links to illustrate the sequence of activities and transactions between roles in the ecosystem. Draw arrows to show the direction of the activity flow. We suggest that you color-code tangible and intangible resources to better see the different types of resources.
5. Review the final business model and validate all the Activities, Resources, Roles, and Sequences with the team. Consider getting additional input and feedback from other people in your firm. Take photos of all whiteboards.

Additional activity resources and templates can be found at www.theartofopportunity.net.

SOURCE: *The Art of Opportunity* authors

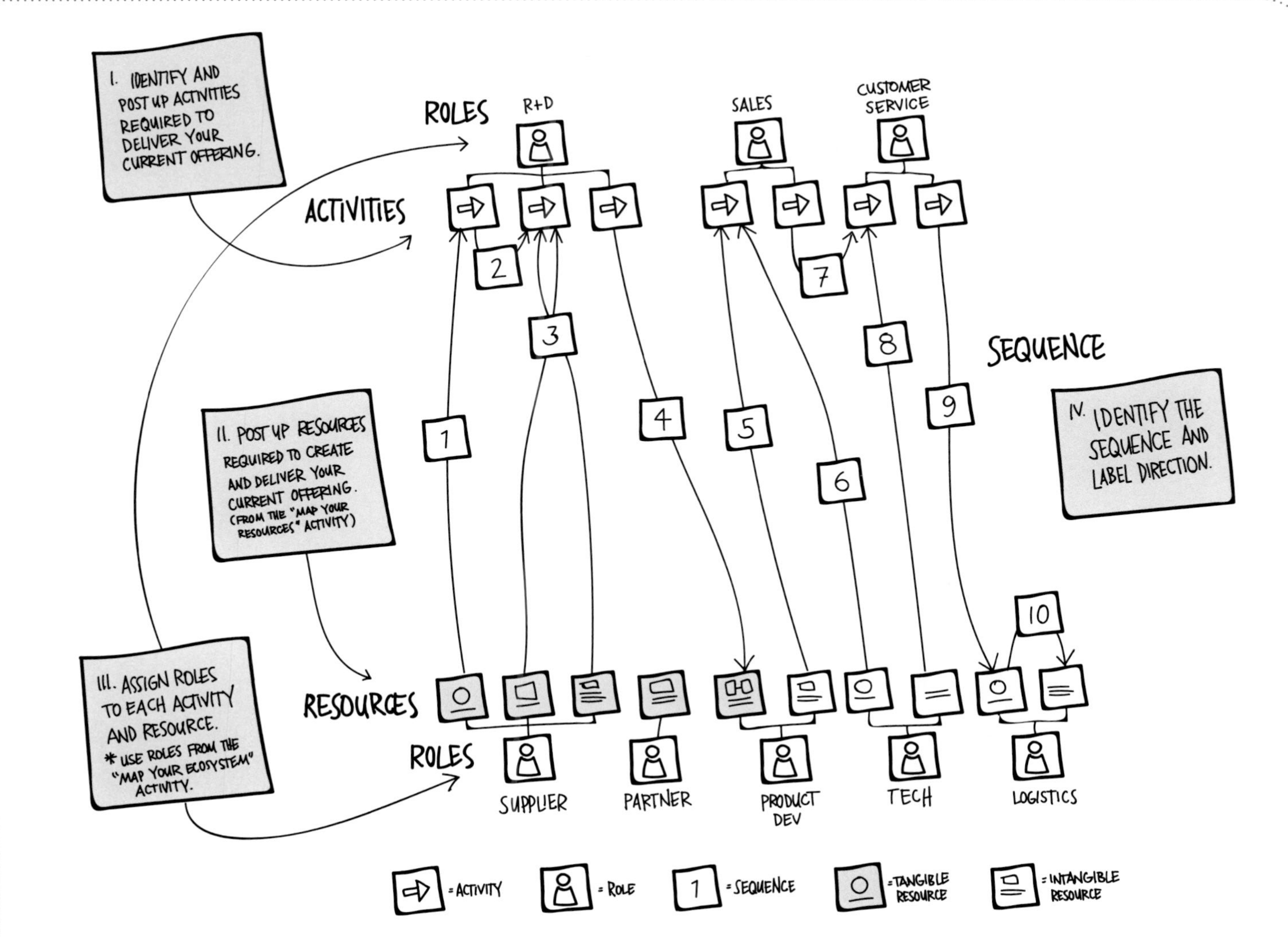
I. IDENTIFY AND POST UP ACTIVITIES REQUIRED TO DELIVER YOUR CURRENT OFFERING.
ROLES
R+D
SALES
CUSTOMER SERVICE
ACTIVITIES
2
3
7
8
SEQUENCE
1
4
5
6
9
II. POST UP RESOURCES REQUIRED TO CREATE AND DELIVER YOUR CURRENT OFFERING. (FROM THE "MAP YOUR RESOURCES" ACTIVITY)
IV. IDENTIFY THE SEQUENCE AND LABEL DIRECTION.
10
III. ASSIGN ROLES TO EACH ACTIVITY AND RESOURCE.
* USE ROLES FROM THE "MAP YOUR ECOSYSTEM" ACTIVITY.
RESOURCES
ROLES
SUPPLIER
PARTNER
PRODUCT DEV
TECH
LOGISTICS
= ACTIVITY
= ROLE
1 = SEQUENCE
= TANGIBLE RESOURCE
= INTANGIBLE RESOURCE

DESIGN YOUR NEW BUSINESS MODEL

Using a set of Business Model Spark Questions, ideate new business model options with your team.

OVERVIEW

Using the visualization of your current business model and Business Model Sparks, generate new ideas for Activities, Resources, Roles, and Sequence to design your new business model.

TIME NEEDED

60–120 minutes

MATERIALS

Markers, sticky notes, whiteboard, previously created visualization of your current business model, and the Business Model Sparks

STEPS

1. Print out the Business Model Sparks and a copy of the visualization of your current business model that you created earlier. Then pass out sticky notes and markers to all participants. Divide participants into teams of three or four. Each team will create a grid on the whiteboard and have rows labeled Activities, Resources, Roles, and Sequence. Write your Breakthrough Question at the top. If you have more than one Breakthrough Question, make a grid for each one.
2. Teams will brainstorm what elements they believe are needed in each of the four business model components to deliver the offering that answers the Breakthrough Question. Begin with the activities required to create and deliver your offering to the customer. Don't be shy about generating a large volume of ideas in each area, and know that this can be a messy and creative process.
3. Next, think about the sequence of your activities. Can you create a new business model by reshuffling it? Try turning the sequence upside down.
4. After having designed a sequence, assign the resources and assets, and roles and capabilities, needed to exercise them, and assign the roles (internally or externally) that will provide the activities. Once you settle on one component idea, be sure to circle back to the other business model components to see if the integrated idea still makes sense.
5. Next grab the set of Business Model Sparks and have each team review them and decide which ones might inspire new ideas to the initially brainstormed business model. After discussion, create a new grid and ideate alternative options for your new business model. Once again, place ideas in each category of Activities, Resources, Roles and Sequence. If you only have minor changes to your original model, the team can choose to just modify the current model.
6. When done, each team shares their ideas with the group. Work to find alignment on the strengths and weaknesses in the business model ideas presented. Then use a decision-making process to select the top one or two business models to explore. Take photos of all whiteboards.

SOURCE: *The Art of Opportunity authors*

Additional activity resources and templates can be found at www.theartofopportunity.net.

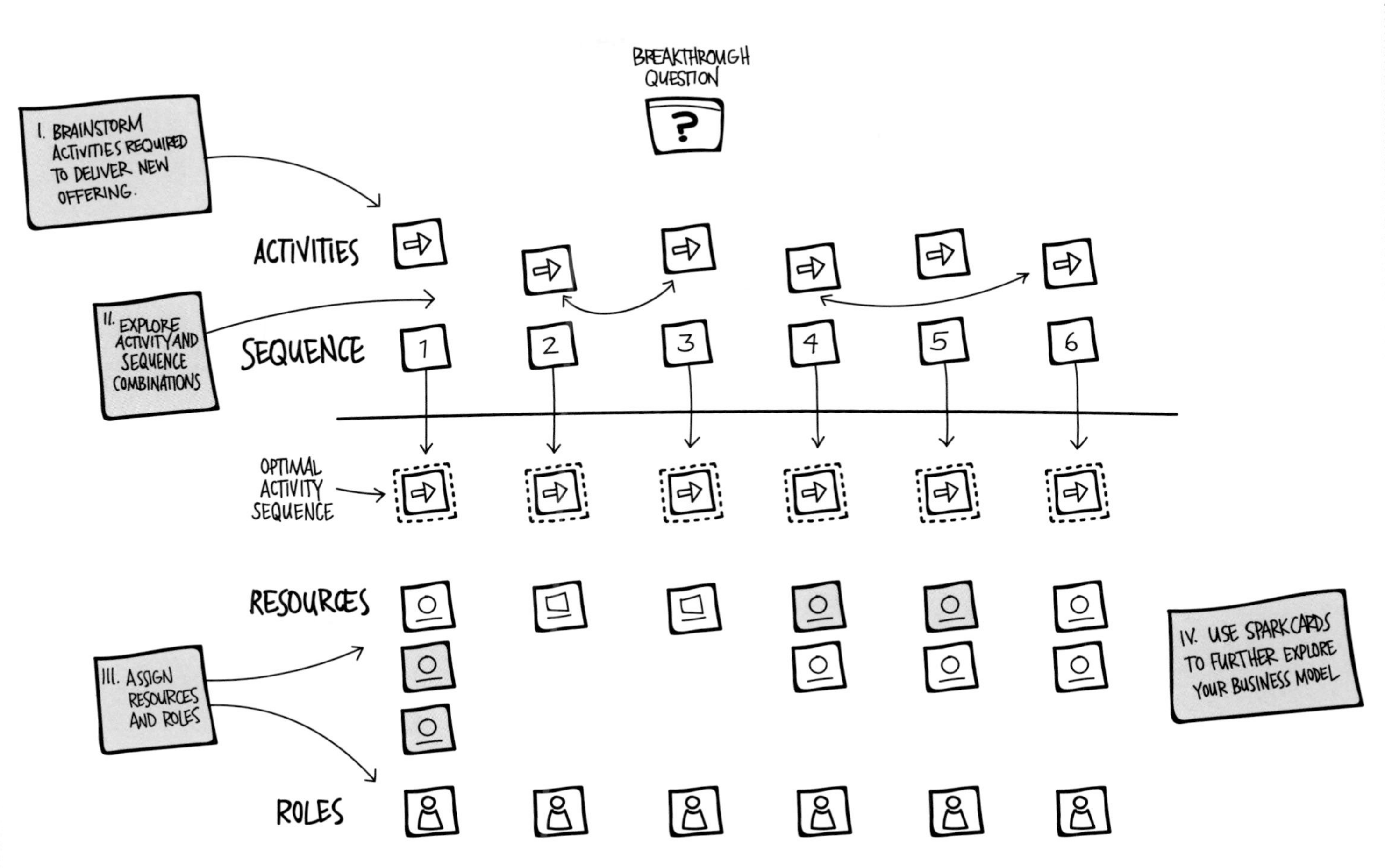
BREAKTHROUGH QUESTION
?
I. BRAINSTORM ACTIVITIES REQUIRED TO DELIVER NEW OFFERING.
ACTIVITIES
II. EXPLORE ACTIVITY AND SEQUENCE COMBINATIONS
SEQUENCE
1
2
3
4
5
6
OPTIMAL ACTIVITY SEQUENCE
RESOURCES
III. ASSIGN RESOURCES AND ROLES
ROLES
IV. USE SPARKCARDS TO FURTHER EXPLORE YOUR BUSINESS MODEL

FLIP YOUR BUSINESS MODEL ASSUMPTIONS

Find new ideas and insights by challenging existing ways of thinking and your core beliefs by looking at completely different or opposite views.

OVERVIEW

Flipping your business model assumptions encourages your team to find innovation alternatives and solutions by questioning the business model assumptions made when designing their new business originally.

TIME NEEDED

60–120 minutes

MATERIALS

Markers, sticky notes, whiteboard, and Design Your New Business Model activity session output.

STEPS

1. Create a grid on the whiteboard with the headers of "Activities," "Resources," "Sequence," and "Roles." In the two rows below the headers, write "Current Assumptions" and "Flipped Assumptions." Give each participant sticky notes and markers.
2. Have each participant carefully review the business model components, and their myriad of elements from your new business model session. Have a team discussion around the set of assumptions you believe to be true that justify the each of the components.
3. In the Current Assumptions row, have each participant write his or her ideas on the key assumptions upon which your business model is based in each of the component areas. Have a conversation, discuss everyone's ideas, and then settle on the key current assumptions.
4. Now for some fun! Have participants create a list of the opposite of each Current Assumption in the Flipped Assumption row. When done, discuss everyone's ideas, and then settle on the key flipped assumption in each component cell.
5. Give everyone plenty of quiet time to write down as many ideas as they can, to reframe the business model based on the Flipped Assumptions. Imagine and describe what would have to be true to make the flipped assumption work. Don't be too hasty in dismissing flipped assumptions. Then, ask each participant to share his or her ideas, giving everyone an opportunity to contribute.
6. Once everyone has shared their ideas, have a discussion to further develop and refine peoples' ideas, and use them to create new ideas. The goal is to gain a different perspective of your new business model, based on seeing it from a different set of assumptions, to find innovative ways to evolve the initial new business model you developed. Go back to your model and revise if you found some brilliant new ideas. Take photos of all whiteboards.

Additional activity resources and templates can be found at www.theartofopportunity.net.

SOURCE: *The Art of Opportunity* authors

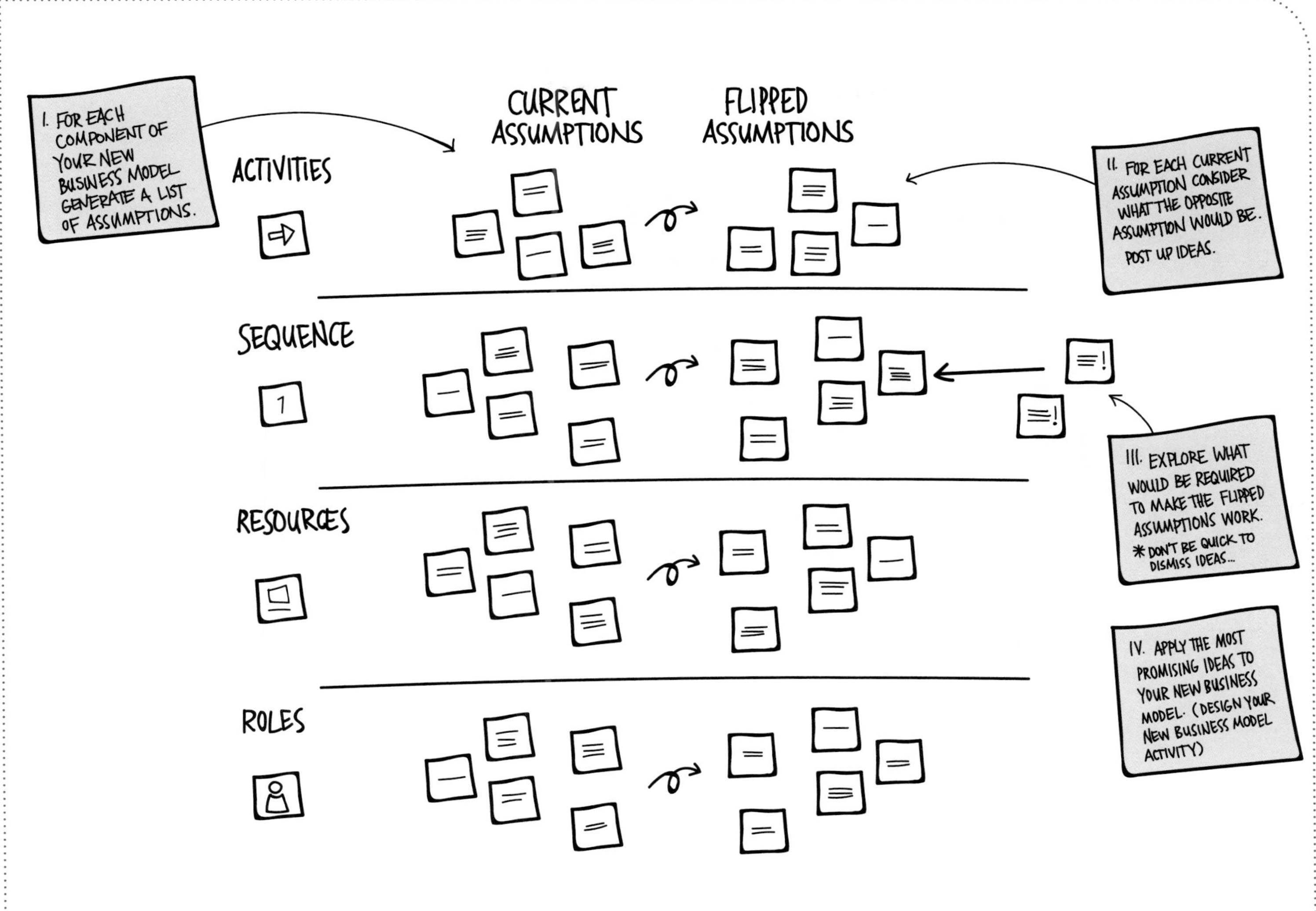
I. FOR EACH COMPONENT OF YOUR NEW BUSINESS MODEL GENERATE A LIST OF ASSUMPTIONS.
CURRENT ASSUMPTIONS
FLIPPED ASSUMPTIONS
II. FOR EACH CURRENT ASSUMPTION CONSIDER WHAT THE OPPOSITE ASSUMPTION WOULD BE. POST UP IDEAS.
ACTIVITIES
SEQUENCE
RESOURCES
ROLES
III. EXPLORE WHAT WOULD BE REQUIRED TO MAKE THE FLIPPED ASSUMPTIONS WORK.
* DON'T BE QUICK TO DISMISS IDEAS...
IV. APPLY THE MOST PROMISING IDEAS TO YOUR NEW BUSINESS MODEL. (DESIGN YOUR NEW BUSINESS MODEL ACTIVITY)

REMIND PEOPLE THAT PROFIT IS THE DIFFERENCE BETWEEN REVENUE AND EXPENSE. THIS MAKES YOU LOOK SMART.

—**SCOTT ADAMS** *Cartoonist*

STRUCTURE YOUR REVENUE MODEL

WHILE DETERMINING YOUR OFFERING and setting your business model are important elements of your strategy, new growth can also emerge through innovative revenue models. Your revenue model is determined by three components:

1. Revenue streams
2. Pricing mechanisms
3. Payment mechanisms

The example of Hilti outlined earlier illustrates how a new business model also required a shift in the revenue model. Hilti's fleet management service turned purchasing products, and hence paying only once when the purchase is made, into a service with recurring monthly payments.

STRUCTURE YOUR REVENUE MODEL

REVENUE STREAMS

Where does the money come from? What is being paid for? Revenue streams are the paths through which money comes to your company or, in other words, **what** customers pay for.

Often there are multiple users who don't have to be the payer. Consider online search models, which offer free search services to users but paid services for advertisers, or TV stations offering free movies, paid for by advertising.

Freemium revenue models are built on the logic that a basic version of the product or service is available for free, whereas premium features need to be paid for. The American rock band, Nine Inch Nails, very successfully applied this strategy when it launched its *Ghosts I–IV* album. The album generated revenues of $1.6 million within the first week of its release. While this might not seem very spectacular, one has to understand that Nine Inch Nails did not work with a record company. The band split with their label in 2007, just before launching the album. And they did not sell the album on iTunes, either. The album was, in fact, available as a free download on Nine Inch Nails' website.

How did Nine Inch Nails make $1.6 million? By offering premium products like special edition CDs, and deluxe and ultradeluxe limited edition packages.

Another example of an innovative revenue model in a well-established industry was set by two coffee shop chains, the French AntiCafé and the Russian Ziferblat. Neither business charges customers for the food and beverages they consume, but rather for the time they spend in the coffee shops. Coffee and food are actually for free. AntiCafé offers hourly, daily, weekly, and even monthly packages of time spent at one of its shops. If you think about it, it's actually logical that in cities like Paris and London, rent is the main cost driver for a coffee shop and not the coffee sold. So why should customers not pay for having access to and using the most valuable asset?

Fractional ownership models for private jets or real estate offered by hotels like the Four Seasons, Fairmont, Ritz-Carlton, and Hyatt bring in additional revenues while granting customers access to a multitude of properties. And, in addition, the customers might profit from the value increase of the properties in which they invested.

Subscription- and membership-based models entitle people to access and usage, not for ownership of goods, and have become widely popular for digital products, especially all kinds of software, as well as for access to digital offerings like music or video streaming. Blacksocks has even turned buying socks into a subscription model. Businesses like gyms thrive on memberships having been paid for, even though most custumers don't regularly use the services.

New revenue streams can also come from using your assets to launch new products and services. Consider Amazon: key assets of its business model are warehouses and IT infrastructure. Leveraging these assets, Amazon created new products and services like Amazon Web Services (AWS) and Amazon Fulfillment, bringing in new revenues. Deutsche Bahn railways offers a car-sharing service called Flinkster, using the cars to generate three different revenue streams: a monthly subscription, a rental fee based on when and how long you use the car, and putting advertising on the car, hence generating advertising revenue.

Revenue stream options to consider:

- **Sales:** Simply selling your goods. Example: Selling a burger.
- **Subscription:** A flat fee (monthly or annually) for access to your offering. Example: a yearly subscription allowing the buyer to have 5 burgers a week.
- **Tiered Subscription:** Different levels of subscriptions at different price points granting access to different levels of offerings. Example: a basic subscription offering two burgers on weekdays, another subscription offering all week access to the restaurant including weekends and five burgers per week, and a VIP subscription offering fast lane entrance, VIP seating, free drinks, and all you can eat burgers.
- **Membership + Usage:** A one-off membership fee, plus additional charges for using the offering. Example: flat monthly fee to be guaranteed a table, plus having to pay for each burger.
- **Advertising:** Renting out assets and space you own to display advertising. Example: advertisements in the menu, on TV screens, and on cups.
- **Sponsoring:** Very similar to advertising. A third party pays for an event, for example, and gains access to your customers in return, or profits from your image. Example: Coke sponsors an event with free burgers for VIP customers.

STRUCTURE YOUR REVENUE MODEL

- **Fractional ownership:** Instead of selling to only one customer, selling the same good to multiple customers. Example: Fractional ownership of private jets, or vacation homes. Members buying a stack in a private burger club.
- **Licensing/Franchising:** Giving somebody the right to use a brand, patent, or any form of intellectual property. Example: Burger King does not operate any of its restaurants, but only grants independent franchise partners the right to do so. Companies license their brands to be used in other domains.
- **Pay-per-use:** Charge a fee for every time a product or service is used. Example: Instead of paying a monthly fee to go to the gym, you actually only pay when and if you go (to work off all these burgers).
- **Leasing/Rent instead of sell/buy:** Instead of selling your good, and transferring ownership, you only grant access based on a fee (often paid as a subscription). Example: Software as a service (SaaS). Rent out your kitchen for customers to cook themselves.

PRICING MECHANISMS

Pricing mechanisms define how much will be paid for your products and services and how the prices will be set. Traditionally, prices tend to be fixed and are set based on a cost-plus strategy, or compared to what competitors charge. An innovative pricing system lets customers determine what they are willing to pay. For example, Lufthansa lets customers make an offer for upgrades to its Premium Economy Class. The airline and hotel industries have refined offering different prices for different levels of services and features, as well as setting their prices dynamically based on demand. During holidays it is common to pay three to four times the price you would pay for the same room during a period of low demand.

The Blue Ocean Strategy framework proposes setting a price that is affordable to the mass of buyers. The so-called price corridor of the mass is thereby not necessarily determined by prices your competitors charge, but also looks at the prices of alternative and substitute offerings that customers refer to in order to fulfill the same need as your product. Within the price corridor of the mass, upper, medium, and low-level prices are determined by how prone your offering is to be imitated. The easier it is for a competitor to copy your offering, the lower the price should be.

Pricing mechanisms to consider:

- **Fixed list price:** A fixed price set by the seller.
- **Flexible price:** Prices are variable and vary depending on criteria like volume bought, time of year.
- **Product feature dependent:** The more features, the more expensive.
- **Customer characteristic dependent:** Prices are set based on the buyer. For example, loyal and frequent customers get a discount, while overseas customers have to pay a higher price.
- **Volume dependent:** The more one buys, the lower the price per item.
- **Value-based:** Instead of fixing prices based on cost and targeted profit margin, the price is based on the value an offering has for the buyer.
- **Bargaining:** Prices are negotiated between seller and buyer.
- **Yield management/demand-based pricing:** If a good is in high-demand, prices go up; if there is low demand, prices go down.
- **Auction:** Buyers compete; the highest bidder receives the good.
- **Reverse auction:** The seller announces a price and if no one buys, the price is lowered until a buyer accepts.
- **Pay what you want/User-defined pricing:** The buyer can pay whatever she wants, and the seller accepts unconditionally.
- **Flat rate:** The all you can eat buffet.
- **Performance-based:** The seller agrees on achieving a certain target, and is paid only if he achieves it.
- **Freemium:** A basic version of the offering is for free; additional features and advanced versions have to be paid for.
- **Premium:** The price is not determined by actual cost, but is influenced by brand, image, exclusivity, etc.

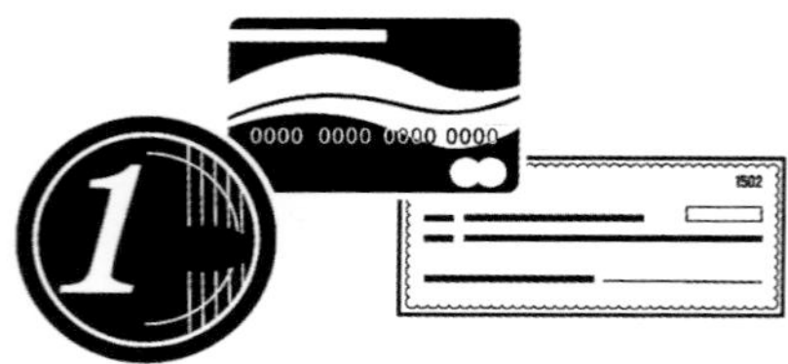

PAYMENT MECHANISMS

Payment mechanisms determine when, how often, and in which form you will be paid.

Payment timings vary from paying in advance to paying after you have received the goods or services. Or one can receive payment through installments or based on milestone completion. Frequency can go from paying once to recurring payments. Payment forms include cash, credit card, wire transfer, bank transfer, and services like PayPal, Amazon Payments, or Bitcoin, as well as whether products and services might be financed by your company. Automotive companies, for example, often offer financial services.

The German-based Saturn and Media Markt electronics stores, for example, offer zero percent financing for their products, enabling customers to buy goods and pay for them in monthly installments, rather then having to pay up front.

Here are some payment mechanisms you can consider:

TIMING:

- **Up front:** The buyer has to pay before receiving the goods/services.
- **After:** The buyer pays after having received the goods/services.
- **Installments:** Instead of paying the full amount at once, the buyer pays in monthly installments, usually after having received the goods.
- **Subscription/Membership:** A flat monthly or yearly fee.
- **Revenue before cost:** Making buyers pay, before the seller has actually had to spend money on the good/service.

PAYMENT METHODS:

- Cash
- Credit cards
- PayPal
- Amazon Payments
- Bitcoin
- Wire transfer
- Check
- Bank transfer

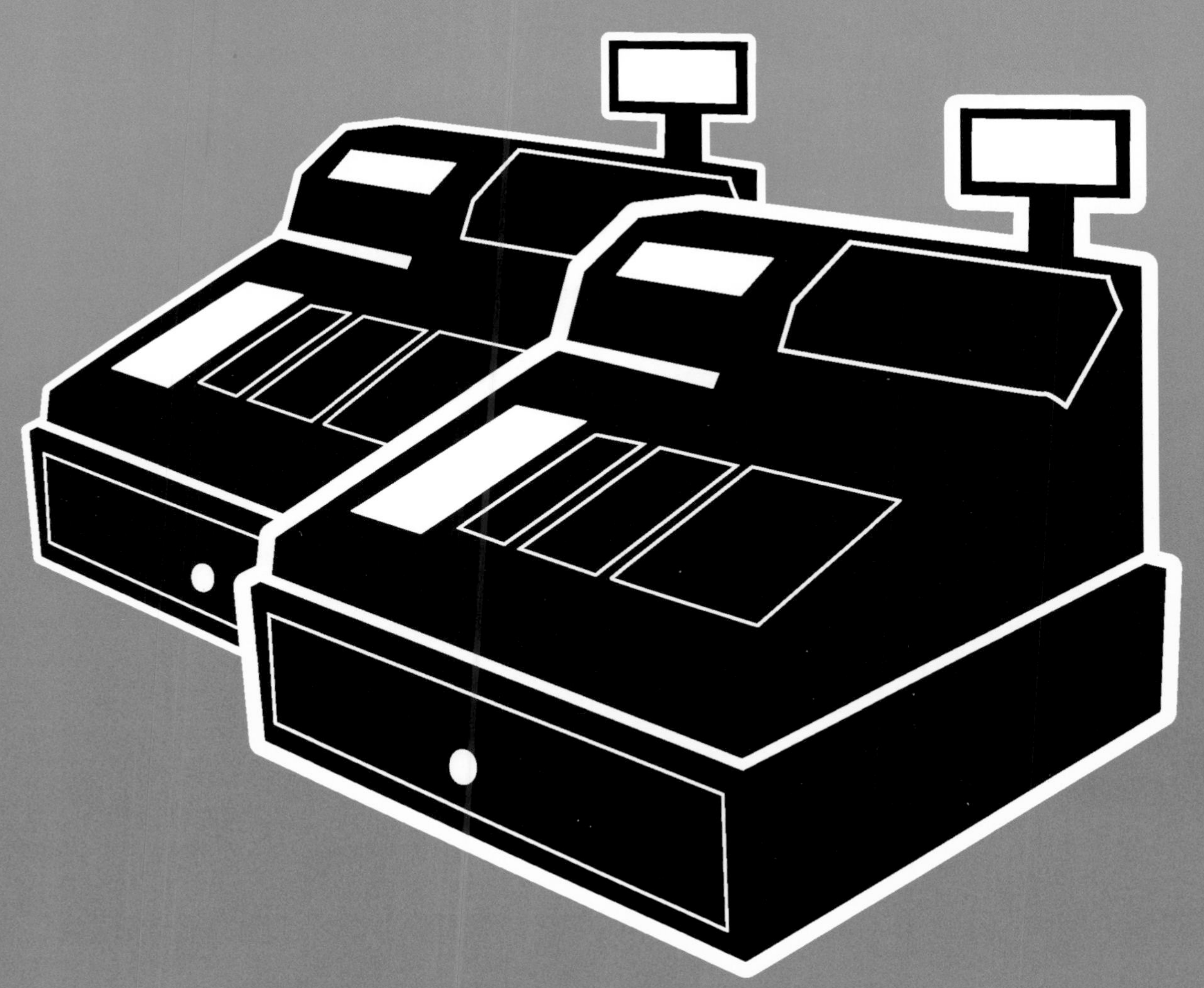

INSPIRATION

PROSIEBENSAT.1 FINDING A REVENUE MODEL FOR THE NONCUSTOMER

The new business of ProSiebenSat.1 (Pro7), which we introduced in Chapter 1, is built on a completely new revenue model, enabling Pro7 to offer TV advertising to customers who have traditionally been ignored by the industry. The typical TV advertising customers are large multinational companies, which buy large amounts of advertising minutes, usually at heavy discounts. Start-ups and small and medium-sized companies firms cannot afford to pay the list prices or buy the high volumes that the larger multinationals buy.

Pro7 created a new revenue model consisting of a base fee, with additional options for a media-for-revenue-share and a media-for-equity component. As such, companies do not have to pay directly for the advertising, but only pay a share of the revenues generated by their media. In the case of start-ups and the media-for-equity model, TV media is used to boost company sales and Pro7 benefits from equity proceeds and in the case of an initial public offering.

The revenue model created a new revenue stream from former noncustomers, and enabled Pro7 to make a profitable offering available to these customers. Pro7 took the new revenue model even further and actually developed a new business model, strategically investing in selected companies and businesses, building a portfolio of investments and a leading position in areas such as online travel businesses, home and living, and fashion and lifestyle.

PRO7 TOOK THE NEW REVENUE MODEL EVEN FURTHER AND ACTUALLY DEVELOPED A NEW BUSINESS MODEL.

INSPIRATION

FAHRENHEIT 212 INNOVATING THE REVENUE MODEL

The New York–based innovation consultancy Fahrenheit 212 has created a business around an innovative revenue model. Typically, revenue models in the consulting industry are built on customers paying for the amount of consulting hours consumed. Sometimes they might pay on a project basis, but project costs are again calculated on the number of days necessary to complete the project.

Fahrenheit 212 offers its customers a revenue model whereby two-thirds of the consulting fee depends on Fahrenheit 212 meeting milestones in the innovation process. Payment is hence spread over multiple stages, and the customers pay only once a milestone is met, thereby reducing their risk in having to pay for a project that does not deliver. Fahrenheit 212 ties a portion of the two-thirds to the commercial success its innovations achieve in the market, creating an additional revenue stream beyond its project fees. And, having gained considerable experience with innovation projects, it is able to charge higher prices now than when it began its business, setting prices more often on the value it creates, than on a cost- or margin-based calculation.

826 VALENCIA MARRYING PROGRAMS FOR NEW REVENUE

The San Francisco–based nonprofit organization, 826 Valencia, supports students between ages 6 and 18 to improve their writing skills. What started with after-school tutoring has developed into a range of programs (e.g., field trips, summer school, workshops, etc.) over the years. The most innovative part of 826 Valencia's business is that the programs are run out of the back of a pirate supply store (yes, pirate supplies!). As the tutoring program is offered for free, 826 Valencia needed a revenue stream. The location and the old store provided the perfect venue for pirate supplies. Meanwhile, other 826 stores around the United States have followed. New York City operates a superhero equipment store, and Los Angeles has its Time Travel Mart.

To learn more about 826 Valencia, listen to the great TED talk on the history of the organization, by its founder, Dave Eggers, when he accepted his 2008 TED Prize.

REVENUE MODEL SPARKS TO CONSIDER

Deciding how to structure your revenue model is a critical step in determining your new business growth strategy. You will need to weave the right blend of revenue streams, and pricing and payment mechanisms to settle on the optimal revenue model. To help you explore options, following is a set of revenue model spark questions to consider. When designing your new revenue model, keep ease of transaction and the satisfaction of your customers and noncustomers in mind.

REVENUE STREAM

- What are our current revenue streams?
- What specific product or services do our customers buy?
- How can we use our assets for additional revenue streams?
- What is the most valuable product or service for our customers or noncustomers?
- What if we gave away our core product for free? How could we still make money? What is the source of that money?

PRICING MECHANISMS

- Is our price affordable to the mass of customers?
- What are the prices of comparable, alternative, and substitute offerings?
- How do we usually set prices—based on cost, benchmarked against competitors, based on performance, based on value created, or other?
- What is the item that defines price?

PAYMENT MECHANISM

- When do we ask customers to pay?
- Do we need the money up front?
- Could we offer installment-based payments?
- How easy or difficult is it to pay us?
- Which payment options would our customers prefer?
- Could we entice customers to pay up front?
- Can we tie the payment to specific events in the buying/consumption process?

REVENUE MODEL CARD SORT

The revenue model card sort helps explore, define, and structure the architecture of your revenue model.

OVERVIEW

Card sorting will help you better understand the implications of employing different options for building your revenue stream, and pricing and payment mechanisms.

TIME NEEDED

60–120 minutes

MATERIALS

Markers, sticky notes, Growth Initiative Brief, Breakthrough Questions, Revenue Model Options, and Revenue Model Sparks

STEPS

1. Prepare by either creating and printing out pages of the options in each of the three revenue model components (revenue stream, and pricing and payment mechanisms) or downloading, printing, and cutting up the revenue model cards from the *Art of Opportunity* site.
2. Take your set of cards and put them into three separate stacks, once for each component (revenue stream, and pricing and payment mechanisms).
3. Divide your group into teams of two or three. Be sure all the cards for each component are visible. Using the Growth Brief as a guide, have participants brainstorm different combinations of options, using one or more cards from each component area. Use the Revenue Model Sparks for inspiration as you create your combinations. Put your ideas on a whiteboard with three columns labeled Revenue Stream, Pricing Mechanisms, and Payment Mechanisms.
4. When done, allow everyone to briefly describe his or her combination ideas and allow time for group discussion. Does each combination overcome certain barriers or hurdles and/or simplify or answer a customer or noncustomer need?
5. Once everyone has shared their ideas, start a group discussion to further refine and develop combinations, and use them to create new ideas. Criteria to consider: best, worst, most innovative, new to the company, new to the industry, new to the world, unique. Take photos of all combinations.

Additional activity resources and templates can be found at www.theartofopportunity.net.

SOURCE: *The Art of Opportunity* authors

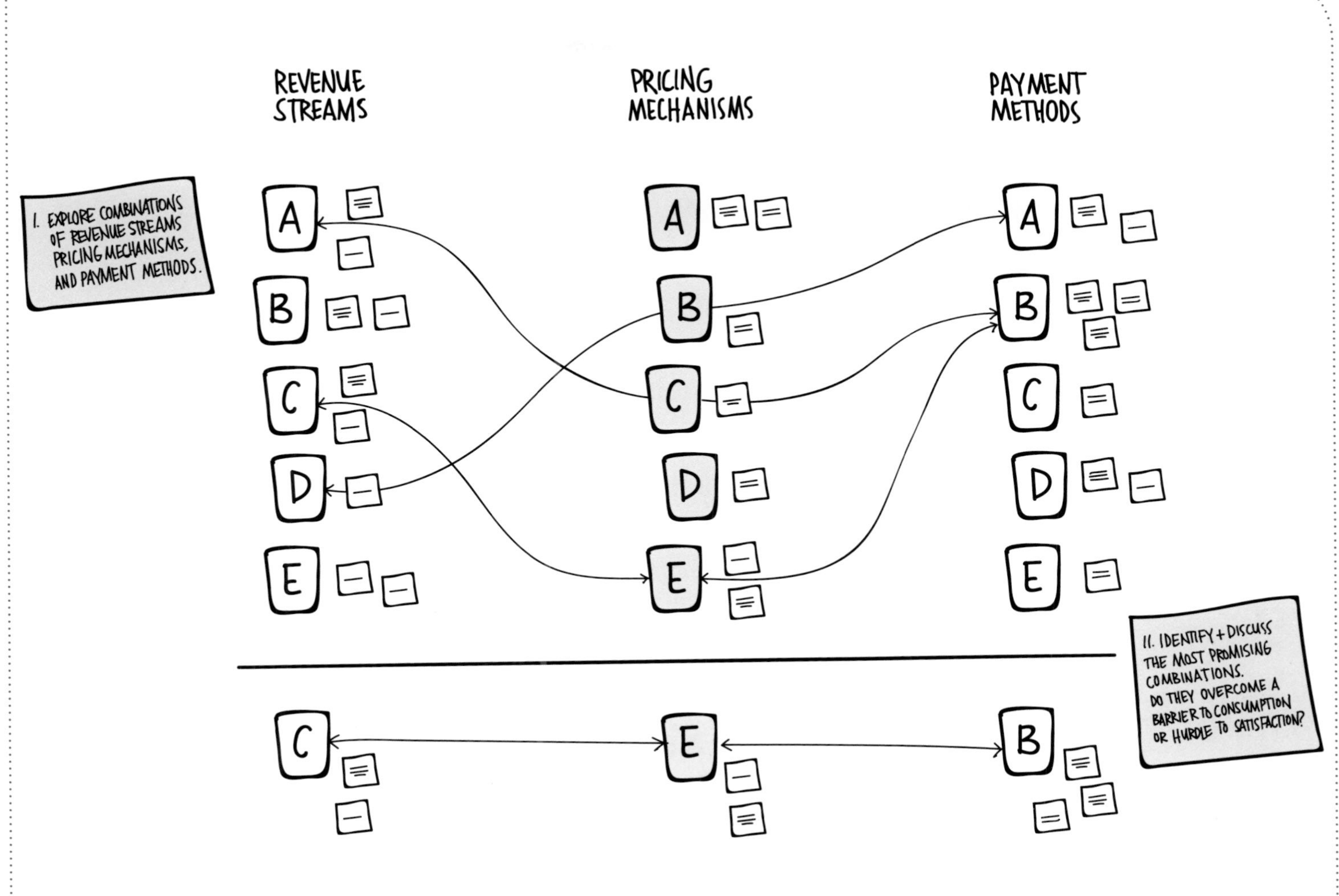
REVENUE STREAMS
PRICING MECHANISMS
PAYMENT METHODS
I. EXPLORE COMBINATIONS OF REVENUE STREAMS PRICING MECHANISMS, AND PAYMENT METHODS.
A
B
C
D
E
II. IDENTIFY + DISCUSS THE MOST PROMISING COMBINATIONS. DO THEY OVERCOME A BARRIER TO CONSUMPTION OR HURDLE TO SATISFACTION?

SET YOUR STRATEGY: CREATING VALUE

YOU'VE MOVED THROUGH the myriad of ways to craft your strategy, exploring your offering, your business model, and your revenue model, having developed multiple options. Now it is time to make some decisions, setting the strategy you believe will be most valuable to build your growth business. How can you assess the potential of success of your strategy options?

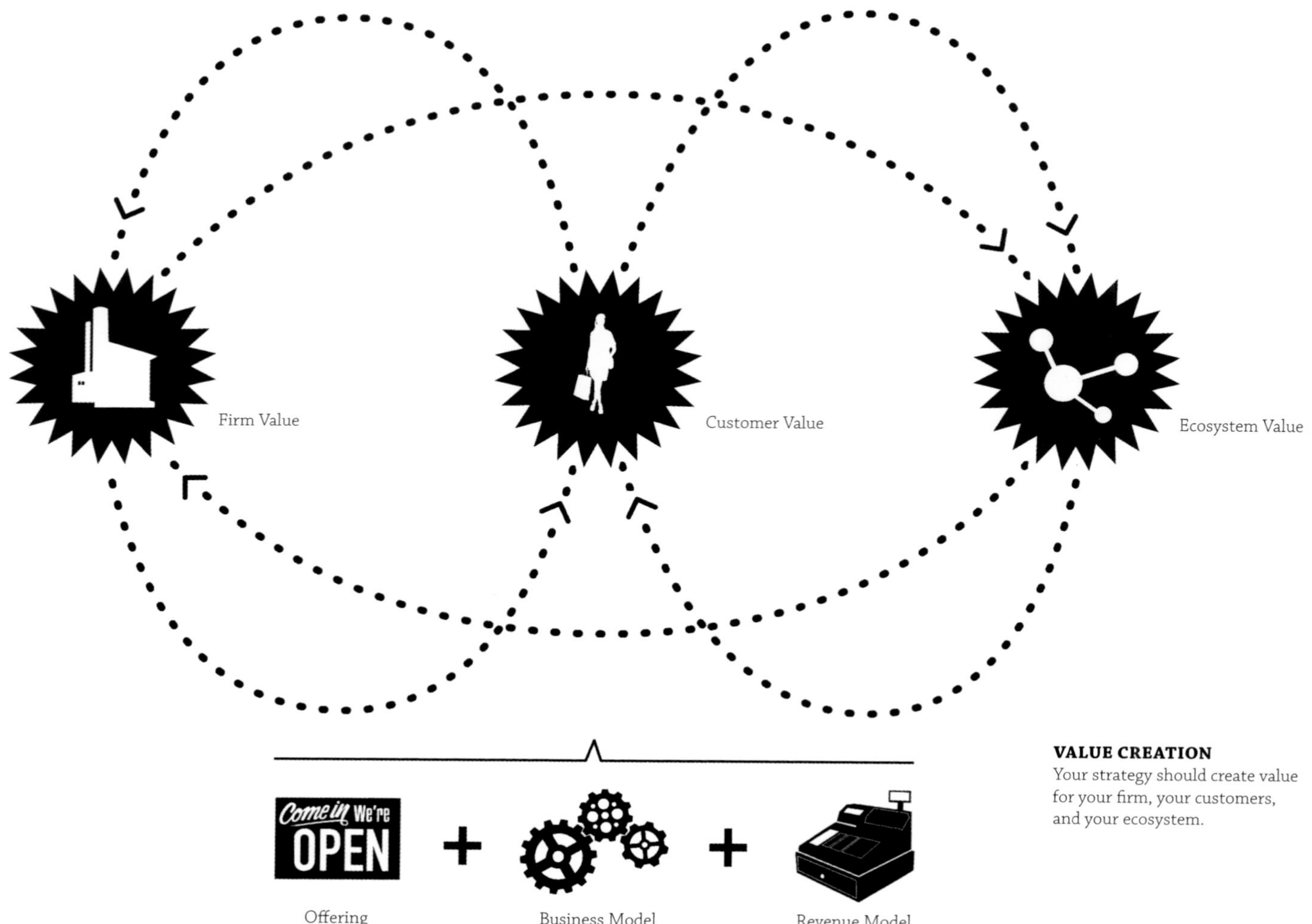

VALUE CREATION
Your strategy should create value for your firm, your customers, and your ecosystem.

CREATING VALUE

THINK ABOUT HOW YOU CAN WIN in today's business environment. Instead of competing by focusing on low cost and/or differentiation, the winners in today's economies focus on creating value and benefits for multiple stakeholders. Your growth strategy should create value for your:

1. Customers
2. Firm
3. Ecosystem

Evaluating optional strategies using the criteria of value and benefit creation helps you to select which strategy to pilot in the market. In practice, the three types of value might not be equally important; yet, if you manage to design a strategy that creates strong benefits for every group, you will achieve much stronger success.

CREATING VALUE

CUSTOMER VALUE

Your growth strategy will be successful only if you create value and benefits for your customers. Creating value for your customers is a function of taking your specified customer segment, along with your identified opportunity (based on the customer's problem, need, and job to be done) and your strategy. Your strategy is the solution you are proposing to solve the customer's need, and consists of your offering, your business model, or your revenue model.

Often your customer might not be the final user of your products or services, or you might have multiple customer groups you need to think about. As shown earlier in the example of Hilti, the firm's fleet management and the logistics and financial administration services are targeted at offering different buyer groups very specific value propositions.

Hilti offers its customers high-quality and reliable tools, and, as mentioned, workers are proud to show their red-painted Hilti tools. For the tool-crib manager, it is easy to order new tools and manage the inventory of existing tools. For the CFO, the monthly subscription-based fleet management service makes it easy to plan cash flow and expenses.

To evaluate whether your strategy creates value and benefits for your customers, go back to the opportunity you discovered and ask whether your strategy removes the barriers to consumption and the hurdles to satisfaction you identified along the customer journey.

CUSTOMER VALUE QUESTIONS

Here are a few questions to use in evaluating how well your strategy solves your customers' needs to create value for them.

- How does this strategy enhance customer productivity?
- How does this strategy solve the need of the customer?
- How does this strategy make the customer journey more convenient, pleasant, and enjoyable?
- How does this strategy reduce the risk for customers? Bear in mind that risks could be financial, physical, social, or emotional.
- How does this strategy make it easy for the customer to do business with us?
- How does our revenue model make the offering more affordable to customers?
- How will engaging in business with us and using this offering make the customer feel? What will be the image customers are likely to have of us?
- How easy or difficult to use is this offering?
- How do this business model and revenue model make the offering available to our customers?

CREATING VALUE

FIRM VALUE

Besides creating value for your customers, your strategy must create value for your firm. These benefits can be financial, operational, and strategic.

The most obvious **financial** value created for your firm is the profits your strategy creates. Other financial benefits and advantages created, especially by the business model, can be a lower need for financing or a reduced cost. For example, the direct sales and distribution model pioneered by Dell, along with how it handles its logistics, enabled the firm to have negative working capital, and hence, not require a lot of financing. Financial benefits include increased performance in terms of revenue growth, lower cost and thus higher profits, reduced capital investment, and the transformation of fixed cost into variable cost.

Operational benefits include higher flexibility and speed in terms of reaction to the market, higher degrees of asset utilization, and faster inventory turnover. Other benefits could include leveraging existing systems for higher efficiency, better utilization of full-time staff, or reduced complexity.

Strategic benefits include competitive advantage through differentiation, a unique position in the market through the creation of superior value for the customer, increased market share, enhanced brand and reputation, and the exploitation of business opportunities. Strategic benefits also include the further possibilities that a certain offering, business model, or revenue model creates for your company. In the case of Amazon, its online business model has enabled the firm to expand from books into all sorts of product areas, a possibility that the traditional bookstore does not offer. The underlying assets and capabilities necessary to operate the Amazon business model have created further opportunities to venture into online web services, logistics, and payment services. For Hilti, fleet management customers are more loyal and satisfied with the brand and, as a result, spend more with Hilti.

FIRM VALUE QUESTIONS

The following are questions to use in evaluating how well your strategy creates value for your firm.

- What strategic, operational, and financial benefits does this strategy create for our firm?
- How does this strategy put us in a position of competitive advantage?
- How difficult or easy will it be for others to imitate this strategy?
- Which further options and opportunities does this strategy open up for us?
- How does this strategy enable us to turn noncustomers into customers?
- How does this strategy leverage our unique assets and capabilities?
- How difficult or easy will it be to implement this strategy?

CREATING VALUE

ECOSYSTEM VALUE

Finally, your business model needs to create value for your ecosystem and the suppliers, partners, and stakeholders. If it doesn't create any value for these ecosystem roles or actors, why should they engage in any business activity with your company? As with firm value, the ecosystem's value can be financial, operational, strategic, or even emotional.

The tutoring offering of 826 Valencia creates value for the students (who get better at their schoolwork), as well as value for the parents (who don't need to spend as much time helping their children to study and are proud of their kids when they receive better grades). The teachers working with 826 Valencia also receive value, as the tutoring supports their efforts in teaching the children, and the tutoring organization creates emotional value for the volunteers who work with the kids. The tutoring organization also creates value for the whole community surrounding the store as people get together and socialize during special events.

ECOSYSTEM VALUE QUESTIONS

These questions can be used to determine how well your strategy creates ecosystem value.

- How does this strategy create value for our ecosystem?
- How does this strategy enable us to establish stronger relationships with our ecosystem?
- How does this strategy leverage our unique ecosystem?
- How does this strategy help our ecosystem partners to achieve their objectives?

NEW

INSPIRATION SUMMARY

The following is a summary of some of the examples that have been discussed so far in this book that illustrate the main value and benefits created by a blend of strategies. To see the full list of Inspiration cases, go online to the *Art of Opportunity* site at www.theartofopportunity.net.

VALUE PROPOSITIONS BY COMPANY

	CUSTOMER VALUE PROPOSITION
826 VALENCIA	Supporting kids to improve their writing; homework gets done
AMAZON	Convenience; multitude of products and services from a single source
PROSIEBEN-SAT.1	Access to media for start-ups; no up-front cash expense
EDEN MCCALLUM	Customization of offering and team; lower-cost, high-quality, experienced consultants
KLINIK HIRSLANDEN	High-quality medical and nonmedical services; excellent quality; access to broad range of treatments

FIRM VALUE PROPOSITION	ECOSYSTEM VALUE PROPOSITION
Doing well	Support parents and teachers, as kids get support with their homework and improve their skills; volunteers contribute to their communities
Potential to expand from online book sales into other product and service areas	Delivery partners garner new revenue streams; new partners are developed through evolving business services
Access to new customer segment; profiting from increased sales and equity proceeds	New partners are developed through evolving business services
Low overhead cost; flexible access to talent as needed	Consultants: Flexible work time; opportunity to decide which projects to work on; financially interesting; no need for business development or administrative work
Possibility of providing excellent medicine; lower overhead as doctors are not employed; reputation	Private practitioners: Access to high-quality infrastructure and services; peace of mind as patients are provided with full range of services and care; being part of a strong brand

SET YOUR STRATEGY

Leverage understanding and input from your team to evaluate your growth strategy.

OVERVIEW

Using the criteria of value creation for all of your key stakeholders, this activity enables each team member's voice to be heard in assessing the strength of your strategy.

TIME NEEDED

60–120 minutes

MATERIALS

Markers, sticky notes, whiteboard, Growth Initiative Brief, output from your previous Offering, Business Model, and Revenue Model activity sessions

STEPS

1. Having set your strategy, you will work with your team to assess how well your strategy meets the criteria of creating value for your customer, firm, and ecosystem. Give each participant sticky notes, markers, and a copy of your Growth Initiative Brief and new business strategy. Assign one team member as the scribe. Begin by having participants review the goals of your Growth Initiative Brief, your identified customer, and your Breakthrough Question.
2. Write your Breakthrough Question at the top of a large whiteboard. Then draw 6 columns, labeled: Stakeholder, Opportunity, Offering, Business Model, Revenue Model, and Totals. Create four rows below the column labels. Label the rows Stakeholder: Customer, Firm, Ecosystem, and Total. Finally, hand out a set of the strategy value questions from the book to help explore the value created for each of the three stakeholder groups. There are questions for each group: Customer, Firm, Ecosystem.
3. Have each team member privately assign a numerical value to each cell, using a 1–5 rating, with 5 representing the optimal degree to which a strategy component delivered value for a given audience. Have participants post their value scores, then have the scribe tally the total combined score and create the average score for each cell, and for the total of each column and row.
4. Look across each row and down each column. If any cell has a score of 3.5 or less, you should consider reevaluating that aspect of your strategy.
5. A way to explore specific areas that need to be addressed is to repeat the rating process, but go a level deeper and rate how well the strategy addressed feasibility, desirability, and viability for each audience cell. If any cell has a score of 3.5 or less, revisit this topic to ensure your strategy's success. Take photos of all whiteboards.

SOURCE: *The Art of Opportunity* authors

Additional activity resources and templates can be found at www.theartofopportunity.net.

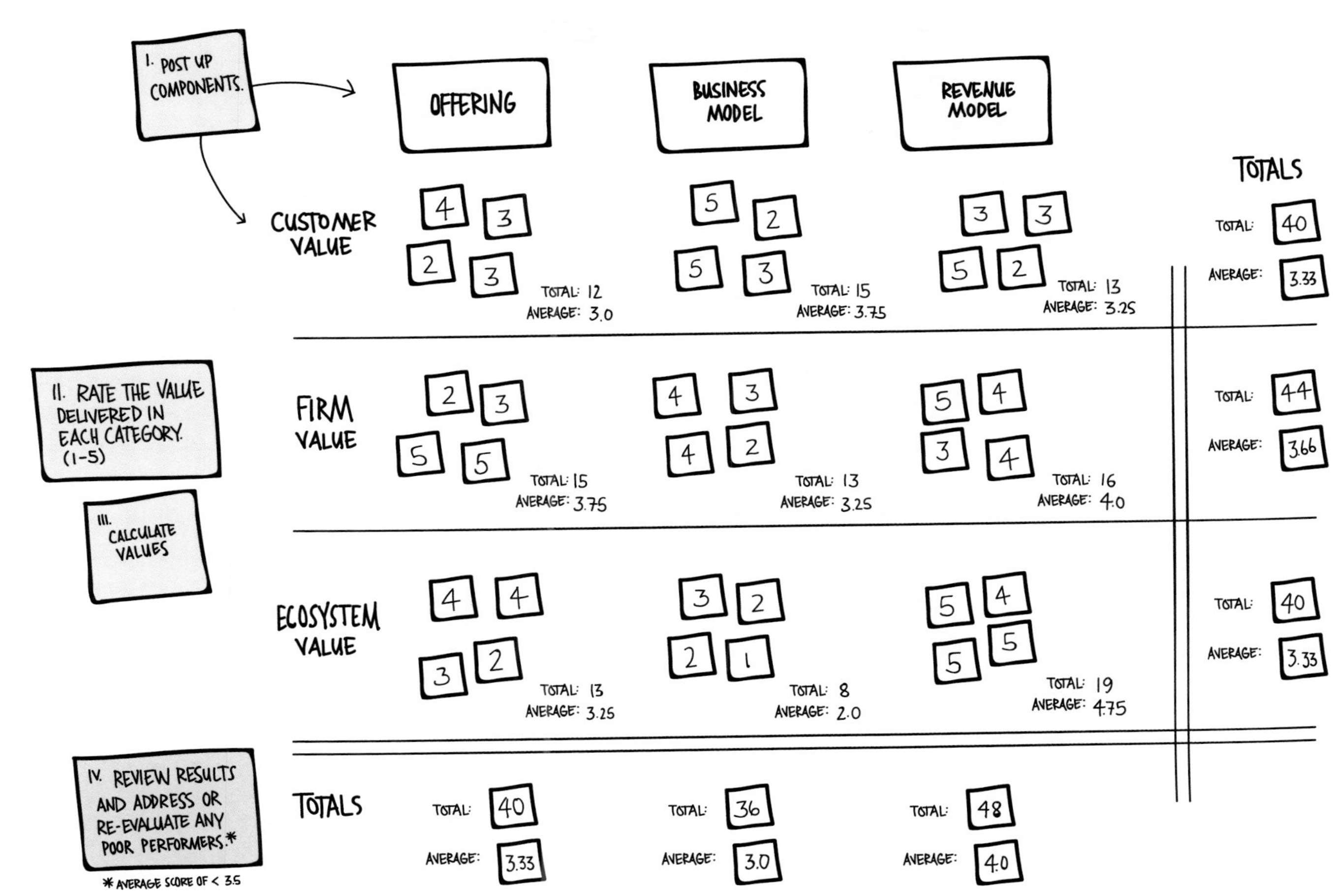
I. POST UP COMPONENTS.
OFFERING
BUSINESS MODEL
REVENUE MODEL
TOTALS
CUSTOMER VALUE
4 3 2 3
TOTAL: 12
AVERAGE: 3.0
5 2 5 3
TOTAL: 15
AVERAGE: 3.75
3 3 5 2
TOTAL: 13
AVERAGE: 3.25
TOTAL: 40
AVERAGE: 3.33
II. RATE THE VALUE DELIVERED IN EACH CATEGORY. (1-5)
FIRM VALUE
2 3 5 5
TOTAL: 15
AVERAGE: 3.75
4 3 4 2
TOTAL: 13
AVERAGE: 3.25
5 4 3 4
TOTAL: 16
AVERAGE: 4.0
TOTAL: 44
AVERAGE: 3.66
III. CALCULATE VALUES
ECOSYSTEM VALUE
4 4 3 2
TOTAL: 13
AVERAGE: 3.25
3 2 2 1
TOTAL: 8
AVERAGE: 2.0
5 4 5 5
TOTAL: 19
AVERAGE: 4.75
TOTAL: 40
AVERAGE: 3.33
IV. REVIEW RESULTS AND ADDRESS OR RE-EVALUATE ANY POOR PERFORMERS.*
* AVERAGE SCORE OF < 3.5
TOTALS
TOTAL: 40
AVERAGE: 3.33
TOTAL: 36
AVERAGE: 3.0
TOTAL: 48
AVERAGE: 4.0

VISUALIZE YOUR STRATEGY

CONGRATULATIONS, you've reached another milestone. You landed on your strategy by working through the three elements of your offering: products, services, and customer experience. Then you shaped your business model from the components of activities, resources, sequence, and roles. Next, mixing the right combination of revenue streams and pricing and payment mechanisms, you shaped your revenue model. Finally, you integrated all those moving parts to assess how they created value for your customer, firm, and ecosystem.

Once again it is time to report on your progress toward launching your new growth business by presenting the strategy you have crafted. Use the Strategy Report template to illustrate the journey you've made and why and how you recommend the strategy you designed. Gather your teammates and work through the template, leveraging all the content you've created in this phase of your opportunity journey.

FINAL NOTE ON CRAFTING YOUR STRATEGY

AS ILLUSTRATED THROUGHOUT THIS CHAPTER, your growth strategy is composed of your offering, your business model, and your revenue model. While various ways to design a new strategy have been presented independently of each other, you have probably noticed that the most innovative strategies take advantage of combining strategy elements to create truly unique businesses.

Also note that the three elements act together in a highly interdependent fashion, as the book's examples demonstrate. So, you will see that changing the business model requires changing the revenue model, which in turn leads to a new offering.

VISUALIZE YOUR STRATEGY

Create the story of crafting your strategy using the Strategy Report Template.

OVERVIEW

Sharing your progress and getting feedback is an important step in your opportunity journey. Using the Strategy Report Template, you can more effectively tell your story, get input, and evolve your thinking before moving to the next phase of building your growth business.

TIME NEEDED

3–4 hours

MATERIALS

Markers, sticky notes, whiteboard, and Set Your Strategy activity output

STEPS

1. Bring your Opportunity Report, and all of your materials from the Strategy phase: outputs from your brainstorms and workshops for your offering, business model, revenue model, and value creation. Print out the Strategy Report Template, or download it from our site.
2. Work through each phase to tell the story of discovering your opportunity. Begin with the headline—Your Breakthrough Question, then work your way through the process of how you arrived at this strategy:
   - What did you do to gain an understanding of your offering and how it integrated products, service, and customer experience?
   - How did you identify your business model and the blend of activities, resources, sequence, and roles?
   - Why did you choose the components of your revenue model (revenue stream, and pricing and payment mechanisms)?
3. For each strategy element, describe what is unique in each area. What might you have removed, added, or changed? Begin by writing a headline and short description of each element (no more than three sentences) and place into the Strategy Report Template. Note illustrations, photos, diagrams, and your prior work—anything that can be used to visualize the descriptions.
4. Now take the output from the Set Your Strategy activity, and use it to describe how you have created value for your customers, firm, and ecosystem. Once again, write a headline and short description for the value created for each stakeholder group.
5. When you have completed creating your Strategy Report, practice presenting it to your team or a friendly, receptive audience before giving a more formal presentation. This method of active iteration and practice allows you to evolve and improve your story.

Additional activity resources and templates can be found at www.theartofopportunity.net.

SOURCE: *The Art of Opportunity* authors

STRATEGY REPORT

BREAKTHROUGH QUESTION

HOW MIGHT WE...

CREATE A NEW MARKET AND MAKE TV ADVERTISING AVAILABLE...

FOR THIS AUDIENCE:

...TO SMALL AND MEDIUM-SIZED ENTERPRISES AND START UPS...

BY OVERCOMING:

...LACK OF FINANCING AND CASH AVAILABILITY WITHOUT SACRIFICING OUR MARGINS...

WITH THESE RESOURCES:

...WITH OUR EXISTING RESOURCES (ADVERTISING MINUTES AND MEDIA POWER)

OFFERING

PRODUCT	SERVICE	CUSTOMER EXPERIENCE
TV MEDIA	MEDIA PLANNING DEVELOPMENT + EXECUTION	ONE STOP SHOP

BUSINESS MODEL

ACTIVITIES	SELECT START UPS	PLAN MEDIA	PRODUCTION
RESOURCES			
ROLES	SELECTION TEAM	OPERATIONS TEAM (MANAGE THE MEDIA PROCESS)	RESOURCE TEAMS
SEQUENCE	1	2	3

REVENUE MODEL

PRICING METHOD	REVENUE STREAMS	PAYMENT METHODS
BASE FEE	EXCHANGE MEDIA FOR EQUITY; EXCHANGE MEDIA FOR REVENUE SHARE	

CUSTOMER VALUE

* ACCESS TO MEDIA POWER
* EXECUTION SUPPORT
* MEDIA DRIVES UP SALES AND COMPANY VALUE AT IPO

FIRM VALUE

* ACCESS TO NEW CUSTOMERS
* HIGH POTENTIAL REVENUE FROM SALES AND IPO EXITS
* STRONG INVESTMENT PORTFOLIOS INCREASE MARKET SHARE

ECOSYSTEM VALUE

* PRODUCTION PARTNERSHIPS

THE DIFFICULTY LIES NOT

SO MUCH IN DEVELOPING

NEW IDEAS . . .

AS IN ESCAPING

FROM OLD ONES.

—**JOHN MAYNARD KEYNES** *Economist*

LAUNCH YOUR NEW GROWTH BUSINESS

PLAN. (v.t.)

TO BOTHER ABOUT THE BEST METHOD OF ACCOMPLISHING AN ACCIDENTAL RESULT.

—**AMBROSE BIERCE** *Journalist*

Having discovered your opportunity for new growth (Chapter 2) and designed your first iteration of the strategy for how to seize this opportunity (Chapter 3), it's time to launch and build your new business. Most firms jump right into creating a complex and overly detailed launch plan. Instead, to reduce risk and increase your likelihood of success, we suggest an iterative approach that allows you to validate, learn, and design your business in an iterative, evolving way.

THE THREE PHASES OF A NEW GROWTH BUSINESS

OUR RESEARCH HAS SHOWN that successful strategic innovators go through three phases to build new growth businesses:

1. The **inception** phase, within which an opportunity for new growth is discovered, followed by the development of an initial idea of the strategy for how to seize this opportunity, and the validation of this idea. (This is actually where you are in your journey at the moment! Whatever strategy you have developed until now, consider it a first idea.)
2. After the successful validation of the initial idea, the new business enters the **evolution** phase, during which the offering, business model, and revenue model are being reworked, fine-tuned, and adapted while the new business is being operated.
3. Once the new strategy works sufficiently well, the business enters what we call the **diffusion** phase, during which the focus of activities shifts from designing and crafting the strategy to scaling up the new business.

The core business design thinking principle underlying this process is active iteration, which focuses on learning through active experience. Whereas traditional strategy processes (and actually most processes in management) rely on first analyzing, planning, and then implementing, the inception/evolution/diffusion (IED) process follows an iterative cycle: act, learn, design, act, and so on.

INCEPTION

EVOLUTION

DIFFUSION

Engaging in action while making small steps launches an iterative feedback loop, which enables you to use the knowledge gained through action to further design, adapt, and pivot your strategy. Repeating this process continually evolves your strategy until it reaches a level of maturity in which the strategy works sufficiently well to allow you to move from testing to scaling it up.

WHAT'S THE DIFFERENCE?

Methodologies like scrum or lean start-up are built on similar principles, yet have been applied mostly to software programming and building and launching online products. We illustrate how these principles can be applied to strategy, going beyond the realm of start-ups, online businesses, and software development. Moreover, as our research has shown, the "act–learn–design" cycle is only part of the story (actually the middle part), and it does not continue forever. Once a sufficiently well-working strategy has been crafted, the focus shifts from designing the strategy to scaling up the business.

In addition, while more and more process theories acknowledge the iterative nature of activities, alternating between action and thinking, they nevertheless still presuppose a natural sequence of first analysis, followed by planning, and finally executing the plan. As our case examples illustrate, and as most people naturally know, this is not how things work in real life.

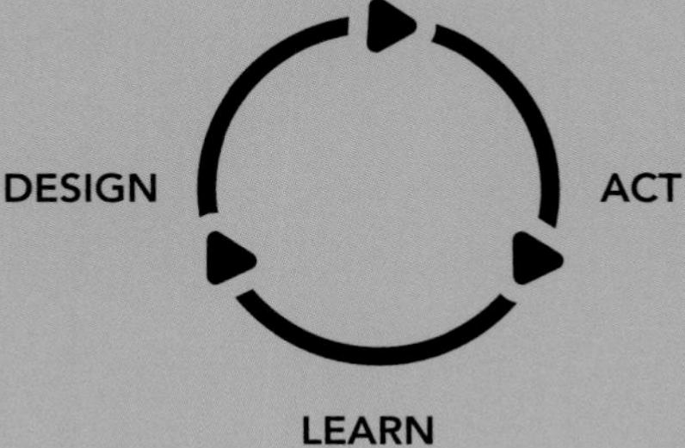

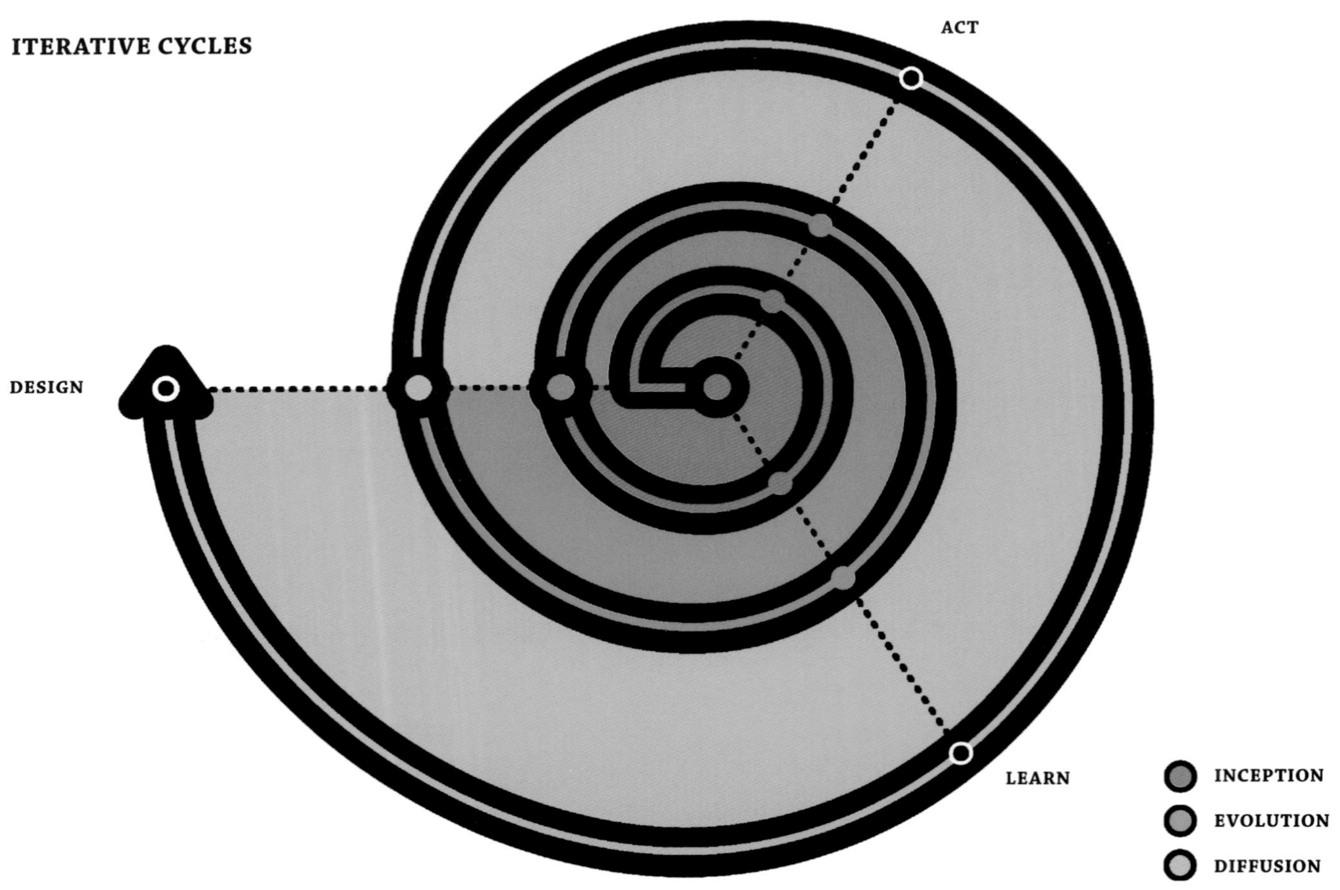
ITERATIVE CYCLES
ACT
DESIGN
LEARN
INCEPTION
EVOLUTION
DIFFUSION

THE INCEPTION PHASE: VALIDATING YOUR OPPORTUNITY AND PILOTING YOUR STRATEGY

HAVING OUTLINED THE STEPS of identifying an opportunity for new growth and designing your initial idea to seize this opportunity in Chapters 2 and 3, we now focus on outlining the final step of the inception phase: validating your opportunity and piloting your strategy.

Assuming you have engaged with customers, noncustomers, and your ecosystem to discover your growth opportunity, you will have developed a sound strategy by now. However, you cannot be 100 percent certain that your new strategy will really work in the market, especially if you are building a completely new business.

Depending on the type of innovation/growth you've identified, you and your company (and potentially the entire world) may encounter multiple unknowns that are not limited to whether your strategy will work, but also include these questions: What will it take to make the strategy work and operate the new business successfully? How do you gain the knowledge and confidence to move ahead? We propose three steps to increase the odds of launching a successful strategy. As you progress through these steps, your knowledge about the feasibility, desirability, and viability of your strategy and new business will increase.

THREE STEPS FOR VALIDATION AND PILOTING

STEP 1: ***Does the strategy make sense, and can we execute it?*** The first level of validation can be thought of as a theoretical or cognitive validation. This step can be carried out internally, discussing and evaluating your strategy with your colleagues, teammates, your boss, and any other internal stakeholders.

At the end of Chapter 3 we presented three types of value propositions, which constitute a first test of whether your new strategy is likely to create value for all parties and hence promises to be successful. Going beyond the value test, you will need to consider whether you can make the strategy work.

Feasibility Questions

- Can you afford the required investment of money, resources, and time to make the idea work?
- Which skills, capabilities, and assets are necessary?
- Do you possess the necessary skills, capabilities, and assets, or do you need to build or acquire them first?
- Do you have access to the needed channels and partners? If not, what will it take to get access, and can you make it work?
- How will you implement this strategy? Which steps will be required?
- Which hurdles and challenges do you expect?
- Will you be able to solve these hurdles and challenges?
- How will you go about achieving this strategy?

Asking questions like these will give you a first indication of how much effort will be necessary to build your new business. In case you don't have the necessary skills, resources, and so forth and the effort of obtaining them seems insurmountable, you might need to adapt your strategy in the early stages of launching your new growth business.

Beyond the purely operational questions, you might also need to incorporate strategic considerations. For example, how well does the new strategy and business fit with your company's overall strategic aspirations? Will the new growth initiative expand or defend a current business? Will the initiative lay the foundations for a potential new business?

To figure out whether your strategy makes sense, it is worth thinking about its underlying assumptions. In other words, ask yourself: What needs to be true for this to work? Then check whether each of these assumptions is valid and can be confirmed.

In some companies you might need to perform financial analyses such as cash payback, return on investment, or net present value and establish a business case. While we believe that it is hard to reliably establish these for new businesses, the effort will at worst give you a general indication of the value of pursuing the idea. If the maximum estimated potential falls below the required effort, it doesn't make much sense to engage in further steps. If you choose or need to engage in such formal planning activities, remember that pretty much everything within such a plan is likely to be an assumption that will need to be validated.

THE INCEPTION PHASE THREE STEPS FOR VALIDATION AND PILOTING

STEP 2: ***Does the strategy resonate with potential customers and partners?***

Once you are convinced you can make the strategy and your new business work, the question is whether it is really interesting to customers. Before launching the new strategy and business in the market, you will want to talk to customers, lay out your offering, business model, and revenue model, or at least the pieces that need to be validated by the customers. Ask them for feedback to gauge the desirability of your new business idea. The first step can be completed internally, but the second requires you to go outside your organization and engage with (potential) customers and partners.

Be sure to provide customers with all the details that might influence their buying decisions along their journey: ease of use, cost, access, etc. Be as specific as possible, and try to get feedback on each element of your new strategy.

For example, if you were to ask customers whether they wanted to fly from London to New York in under four hours, who wouldn't want that? But if you tell them that the cost will likely be around US$18,000 (the approximate cost for a round-trip ticket on the Concorde in 2015 dollars), their interest might fade quickly. (The very last Concorde ticket sold on eBay for about US$60,000 in 2003.)

If your new business revolves around a new product or service, building a prototype can be an effective means to collect feedback. Prototypes can be created many different ways, depending on the type of offering, the target customer audience, and the resources and time available.

If your business model relies on partners, you will need to check with them, too. Are they interested in collaborating with you? Can they perform the necessary activities in a way you need them to for your new business to work, at the required quality level that you and the final customers expect? And can they do all this at a cost that enables the business to be profitable? If your business depends on partners but they are not interested in participating or cannot deliver the products and services at the required cost and quality, you need to rework your strategy. Online shopping for fresh and perishable food has been made easier only recently with logistics companies like UPS or DHL offering food services and operating special refrigerated trucks, for example.

As with the first step, ask yourself what needs to be true for this strategy and business idea to work, and check with (potential) customers and needed partners to determine whether your assumptions hold up. If they don't, rework your strategy. If they do, proceed to the next step.

WHAT DO YOU NEED TO LEARN?

Every strategy and business idea is built on assumptions. The most basic ones are that there is a market and that customers are interested in your offering. The following table highlights common assumptions you should review in order to ensure your success. Ask yourself: What are we trying to test? What are we hoping to learn? What do we need to learn? What do we need to test to move our strategy forward? What can we do immediately, this week, without spending a lot of money and without needing any permission from anybody?

TABLE OF ASSUMPTIONS

OPPORTUNITY	OFFERING	BUSINESS MODEL	REVENUE MODEL	VALUE CREATED
· The opportunity exists. · Who will buy, how much, and why. · The market is sufficiently large to be interesting. · Market growth rates, size, and so on.	· The market is interested in our offering. · We address the major pain points in the customer journey. · Our offering solves the customer's problems.	· We can operate the business. · We can produce at the required quality, in time, and at the required cost level. · We have access to the necessary skills, capabilities, resources, and assets. · We have access to the necessary partners and distribution channels.	· Our margin is sufficiently large to make a profit in line with expectations. · Customers can afford the price we require.	· The strategy will create value for the customers. · Customers value the benefits created. · The strategy will give us strategic, operational, and financial benefits. · The strategy is interesting for our partners and creates strategic, operational, and financial benefits for them.

THE INCEPTION PHASE THREE STEPS FOR VALIDATION AND PILOTING

STEP 3: ***Does the strategy really work in practice?***

Once you have validated your new strategy internally and received positive feedback about your intended offering, business model, and revenue model from potential customers and partners, it is time to engage in experiential action and pilot your opportunity.

We strongly advise you to do a small pilot before committing resources to a full-fledged operation. Such pilots can be relatively easy and do not need to cost a lot of money.

For instance, Eden McCallum developed a creative way to prototype their business idea, by simply posting an ad in *The Economist* to see whether freelance consultants would be interested in working with the company. After having received about 200 applications, the founders met with 20 consultants and decided to start working with 10 of them. Another example is ProSiebenSat.1. The firm advertised its media-for-revenue-share model and received hundreds of letters from interested companies within two weeks. After screening these, the company picked a handful of promising companies to start working with on this new model. Fahrenheit 212 presented its new innovation process and the risk-and-reward business model to potential clients. One client found the new offering, business, and revenue model so compelling that he bought all of Fahrenheit 212's capacity for six months.

Going outside to actually engage with clients and vendors will yield insights and will help you avoid spending significant resources to get market feedback. Working on a small scale with an easy-to-execute prototype can quickly provide insights and enable you to adjust and retest concepts and assumptions.

To further reduce the risk of your experiments, apply the "safe to try" tactic. For each validation step and experiment, ask yourself whether the action you are about to take has the potential to damage your existing and/or new business. If not, it's "safe to try."

THE MYTH OF EXPERIMENTING WITH VARIOUS STRATEGIES

Experimentation has become a popular approach to test strategies and business models, and learn about what works and what does not. The often implicit assumption is that your company will need to experiment with a range of different strategies to figure out which one works best and is likely to be most successful. Whereas on a corporate level experimenting with a multitude of strategies might be doable, we believe that experimenting with various strategies in parallel on a business unit level or even a division level is highly unrealistic. In fact, our research and experience show that business units and single-division companies usually do not experiment with which strategy to employ, but rather with how to make a chosen strategy work or work better.

Consider Amazon, which made the strategic decision to introduce a new offering, namely allowing customers to sell their used books. Selling used books was a new offering and a way for Amazon to generate new growth. First it tried to sell used books on a dedicated page. When this did not work, Amazon provided each user with a dedicated page. This also did not work sufficiently well, so Amazon started displaying the used books next to the new ones—which finally led to success. The strategy of selling used books did not change. What changed was the way it was implemented.

Another example of experimenting with a selected strategy is the case of Klinik Hirslanden, which we described earlier. Certainly, Hirslanden might have been able to test various business models (e.g., the chief physician model, the private practitioner model, or the hybrid model) at various clinics. However, testing all three models in parallel in a single clinic would have been nearly impossible. Instead, the single clinic picked the hybrid model and tried to make it work, experimenting with different solutions only if the first idea didn't succeed.

THE INCEPTION PHASE

HOW TO VALIDATE YOUR PILOT

ACTIVITIES FOR PRESENTING YOUR VALIDATION DATA

Following is a list of validation testing options, which can be applied at different points based on your desired objectives, whether in your conceptual/theoretically in-house exploration, going outside to customers/stakeholders, getting feedback on possible prototypes or schematic models, or actually piloting to build your business on a small scale during your inception phase.

IN-HOUSE OBSERVATIONS

Explore concepts, gather feedback and gain alignment by speaking with internal stakeholders. This method can be done using interviews or simulation and is excellent for exploring concepts and hypotheses for your opportunity.

FIELD OBSERVATIONS

- **Immersion (being the customer):** Gaining actual firsthand experience in the actual world of the customer and non-customer by participating in every phase of their experience often reveals insights and opportunities otherwise not visible through other means.
- **Customer anthropology (a day-in-the-life):** Using multiple forms of observation of the customer and noncustomer, through photos, note taking, and video that can later be synthesized into observations and insight.

SIMULATION

- **Role-play or bodystorm:** Based on interaction and movement with the body, roleplaying or bodystorming is an improvisational technique of physically experiencing a situation to prototype and derive new ideas.
- **Prototypes:**
  - **Sketch:** A visualization to see how a user interacts with the product or service and what transitions in the experience look like.
  - **Storyboard:** A set of sketches or images arranged in a sequence which outline the scenes of a story.
  - **Wireframes:** A lightweight schematic visual representation of a digital product or service interface used to prioritize information (substance and relationships) over decoration (style).
  - **Model:** Use basic materials to build a rough conceptual version of your physical product to gain feedback.

INTERVIEWS

- **Individuals:** Direct 1:1 inquiry provides a deep and rich view into the behavior, thoughts, and actions of a customer or noncustomer.
- **Groups:** A quick way to learn about a group or community of people to understand overall issues, concerns and behaviors.

THE MYTH OF TRIAL AND ERROR

"Fail fast," "trial and error," and similar notions have become popular postulations when it comes to innovation and have led to the acceptance of failure of products, business models, and strategies. Failure is seen as a natural and unavoidable part of developing successful innovations.

Let's be honest. Nobody likes to fail or be wrong. Imagine announcing at the annual shareholders' meeting that you had sunk a couple of million dollars into unsuccessful strategy trials. Nobody would congratulate you on this result.

While there will be challenges, and while these will have to be solved, our research and experience have led us to believe that it is much more important to "try and succeed." Success is what you should be looking for, not errors and failure. Success and positive experiences are vital to increasing confidence in the strategy and maintaining momentum. Achieving success and communicating this success will help dissolve internal resistance and persuade top management to continue supporting the new strategy and to allocate the needed resources to take the next steps.

Learning what works well will also be necessary to adapt your strategy and help overcome the challenges faced. Test quickly and in small increments, constantly evolving and improving. So, instead of accepting failure and errors, try to find what works well and capitalize on these positive experiences by piloting in small increments that can be quickly evaluated, improved, and evolved.

- **Experts:** Experts can provide more in-depth knowledge and technical information, very useful when you are pressed for time.

RESEARCH

- **Desktop:** Through simple research using a search engine, journals and publications can yield new sources of information and provide new data points for consideration.
- **Data mining:** Massive amounts of data are acquired through multiple sources every second.
- **Expert network community:** Question and explore the opinions from communities of online groups with expertise your subject of inquiry.

SMALL-SCALE EXPERIMENTATION

A good method for evolving your strategy with active iteration is to develop a small-scale version of your product or offering. Or, to gauge interest or demand and/or gain a better understanding of how the product will be used in the field, experiment by floating the value proposition of your business through an ad or communication to prospective customers to elicit responses or volunteers. The goal of your experimentation is to gain insight to help evolve your strategy or implementation plan.

ADDITIONAL RESOURCES TO EXPLORE:

- *The Field Guide to Human-Centered Design;* IDEO
- *Design Thinking for Educators Toolkit;* IDEO
- *101 Design Methods;* Vijay Kumar; John Wiley & Sons; 2013
- *Value Proposition Design;* Osterwalder, Pigneaur, Bernarda, Smith; John Wiley & Sons; 2014
- MindTools.com

WHAT TO TEST

Create hypotheses to test based on the assumptions of your new business strategy.

OVERVIEW

To validate your strategy, create a set of hypotheses that can be measured and then used to develop a methodology for testing.

TIME NEEDED

60–120 minutes

MATERIALS

Markers, voting dots, sticky notes, whiteboard, your Offering and Strategy Reports.

STEPS

1. Give each participant sticky notes and markers. Draw three circles and label them for each stakeholder group: Customers, Firm, and Ecosystem. Break into small group teams. Review the information in your Offering and Strategy Reports.
2. Have a team discussion, exploring the key drivers of value for each stakeholder group. In your Set Your Strategy activity session, you had assessed which combination of offering, business model, and revenue model had the greatest value. Now you want to describe that value—in essence the value proposition—for each stakeholder group. Don't be afraid to generate a number for each group. Post them up around each stakeholder circle.
3. Participants now generate a set of hypotheses for linking each value to a stakeholder group. Take plenty of time for this step. When done, have a discussion to review the set of hypotheses, and aggregate them into themes. Write each hypothesis as clear, simple "We believe . . ." statement.
4. Next, give each participant five dots to use in voting on the hypotheses they feel should be tested. Participants cast their votes all at once and may vote more than once for a single hypothesis if they feel strongly about it. Voting criteria to consider are how critical verifying the hypothesis is for the success of your strategy.
5. Once all the votes are cast, prioritize the hypotheses and choose the ones you will test. Take photos of all whiteboards.

Additional activity resources and templates can be found at www.theartofopportunity.net.

SOURCE: *The Art of Opportunity* authors

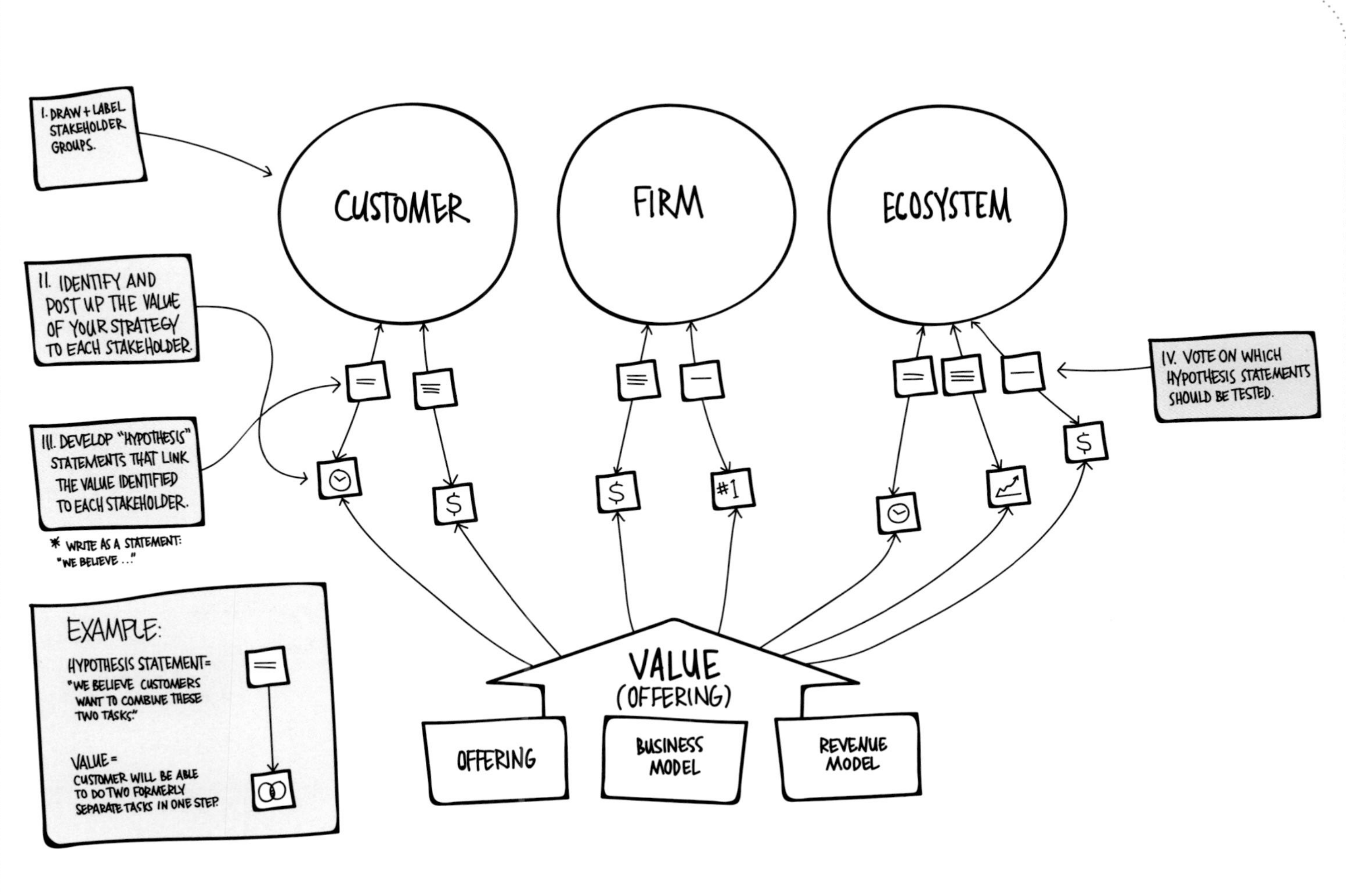
I. DRAW + LABEL STAKEHOLDER GROUPS.
CUSTOMER
FIRM
ECOSYSTEM
II. IDENTIFY AND POST UP THE VALUE OF YOUR STRATEGY TO EACH STAKEHOLDER.
III. DEVELOP "HYPOTHESIS" STATEMENTS THAT LINK THE VALUE IDENTIFIED TO EACH STAKEHOLDER.
* WRITE AS A STATEMENT: "WE BELIEVE . . ."
IV. VOTE ON WHICH HYPOTHESIS STATEMENTS SHOULD BE TESTED.
$
#1
EXAMPLE:
HYPOTHESIS STATEMENT=
"WE BELIEVE CUSTOMERS WANT TO COMBINE THESE TWO TASKS."
VALUE=
CUSTOMER WILL BE ABLE TO DO TWO FORMERLY SEPARATE TASKS IN ONE STEP.
VALUE (OFFERING)
OFFERING
BUSINESS MODEL
REVENUE MODEL

HOW TO TEST

Determine the optimal way to test your hypotheses in order to accelerate learning and evolve your strategy.

OVERVIEW

Choose from a set of validation options to get the best results for evaluation; need, time, and available resources will largely drive your selection.

TIME NEEDED

60–120 minutes

MATERIALS

Markers, sticky notes, whiteboard, Validation Options cards, list of hypotheses from the What to Test activity.

STEPS

1. Print out, or download from book site, the Validation Options cards. Give each participant sticky notes and markers. Review the hypotheses from How to Test workshop. Now have the team identify the audience from whom you should get validation feedback. Do you need a large volume of data and input, or will a few points of reference suffice? Name the "who" to be questioned.
2. Review your list of Validation Option cards and consider the optimal method to get the most accurate results. How much time and what resources are available to test your assumptions? Time and resources will guide with whom and how you can validate and verify your strategy. For example, you decided to create a storyboard prototype to explain a new software platform idea. You determine you will conduct 25 interviews to get user feedback about the value of the platform.
3. For each hypothesis test, have a discussion about how you will measure the results you get from testing and set the threshold for success. Continuing our example, you will measure by administering a brief questionnaire that has a rating system of 1–5 for desirability and adoption. You set your threshold for success as needing to have a 3.5 or higher positive result.
4. Synthesize your ideas into a test plan that outlines your 4 key points: (1) Hypothesis (we believe that . . .), (2) Verified by (name the target test group and what testing method is used), (3) Measured by (define your metrics) and, (4) Success when . . . (the quantifiable way to determine success).
5. Now get testing!

Additional activity resources and templates can be found at www.theartofopportunity.net.

SOURCE: *The Art of Opportunity* authors; Inspiration from *Value Proposition Design*

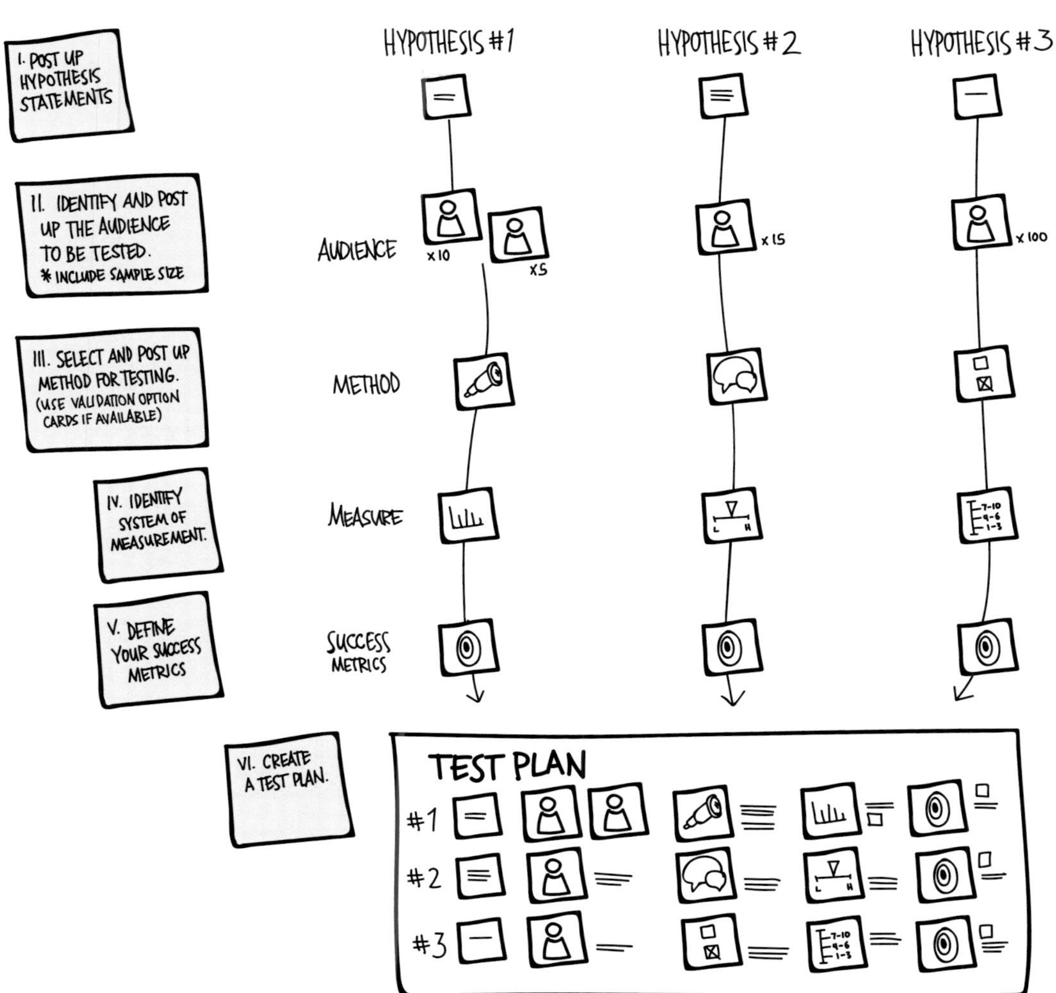

I. POST UP HYPOTHESIS STATEMENTS
II. IDENTIFY AND POST UP THE AUDIENCE TO BE TESTED.
* INCLUDE SAMPLE SIZE
III. SELECT AND POST UP METHOD FOR TESTING.
(USE VALIDATION OPTION CARDS IF AVAILABLE)
IV. IDENTIFY SYSTEM OF MEASUREMENT.
V. DEFINE YOUR SUCCESS METRICS
VI. CREATE A TEST PLAN.
HYPOTHESIS #1
HYPOTHESIS #2
HYPOTHESIS #3
AUDIENCE
x 10
x 5
x 15
x 100
METHOD
MEASURE
SUCCESS METRICS
TEST PLAN
#1
#2
#3

THE INCEPTION PHASE

LEARNING FROM YOUR PILOT

As you engage in action, you will gain a steady stream of information and insight about your offering, business model, and revenue model. The knowledge you gain from your operations will result in three types of experiences:

1. Achieving success
2. Encountering challenges
3. Gaining insights

ACHIEVING SUCCESS

Obviously, the best that can happen is that your pilot demonstrates the success of your strategy. Achieving success means your prototype generates favorable outcomes and the desired results, is accepted by customers and partners, and at the same time is operationally feasible. Acceptance by customers leads to increased demand and financial success.

Achieving success is important for a number of reasons. It legitimizes the new strategy, builds confidence in the idea, wins over skeptics, increases volumes, and provides the opportunity to learn from ongoing experience.

ENCOUNTERING CHALLENGES

While you might have success, at the beginning you are very likely to also encounter challenges. Two types of challenges can be identified at this stage: conceptual and operational.

Conceptual challenges relate to your strategy and suggest that it does not work. Customers do not buy, or they might buy but the revenues generated are not enough for your strategy to be profitable, for example. In short, the strategy does not produce the expected results, whatever they might have been.

Operational challenges relate to the fact that the strategy might work (e.g., customers are interested), but you encounter challenges for customers to be satisfied and for your business to run smoothly. Operational challenges include obstacles such as finding the right skills to run the new business, processes or organizational structures needing to be adapted, partners not performing, or the terms of the collaboration simply not working. Operational challenges often lead to internal resistance, especially if your new strategy is supposed to replace the established one.

GAINING INSIGHTS

No matter what, engaging in business activities to garner experience and feedback will reveal insights that you can apply to tweak your strategy to make it work ever better. You might learn how valuable your services are to customers, which allows you to increase your revenues. Or you learn about particular challenges your customers face, and can address these to enhance the value proposition of your strategy. Or your customers might give you the solution needed to solve a conceptual strategic challenge.

The key is that at a certain point in time, you will need to set some assumptions and create the means to go outside and pilot your strategy, speak with customers, try some operational tests, and engage in action to make the experiences necessary to craft your strategy. The goal is to keep on learning and to create solid data through testing, which will back further decisions.

THE EVOLUTION PHASE: ADAPTING YOUR STRATEGY

ONCE YOU'VE BEGUN TO PILOT YOUR STRATEGY and new business, you will need to adapt and fine-tune them to the types of experiences you encountered and the insights you gained along the way.

You can think of the process of crafting your strategy as active iteration, a series of cycles of engaging in action, reflecting and designing based on experiences made, and engaging in further action. As you repeat cycle after cycle of making experiences, designing, and engaging in action, the maturity level and sophistication of your new strategy will increase.

Crafting your strategy through such iterative cycles liberates you from tedious long-term planning, forecasting, and the burden of having to be right the first time around and having to know all the answers before you start. Only through making experiences will you be able to decide on the right path to follow. Thus, "crafting your strategy" means that you will design your business while operating it, and simultaneously operate it while it is still being designed.

HOW TO FURTHER DESIGN YOUR STRATEGY

Fine-tuning your strategy requires you to go back to your offering, business model, revenue model, and the value propositions, and review each element regarding how well it works in practice.

Initially, we suggest you do reviews and iteration at regular intervals. Depending on the nature of your business, these can be monthly, quarterly, or yearly reviews, for example. These regularly scheduled strategy reviews are necessary to adapt and further design your strategy to make it work even better. Managers often practice some individual self-reflection before engaging with their colleagues and teams to evaluate the current state of affairs.

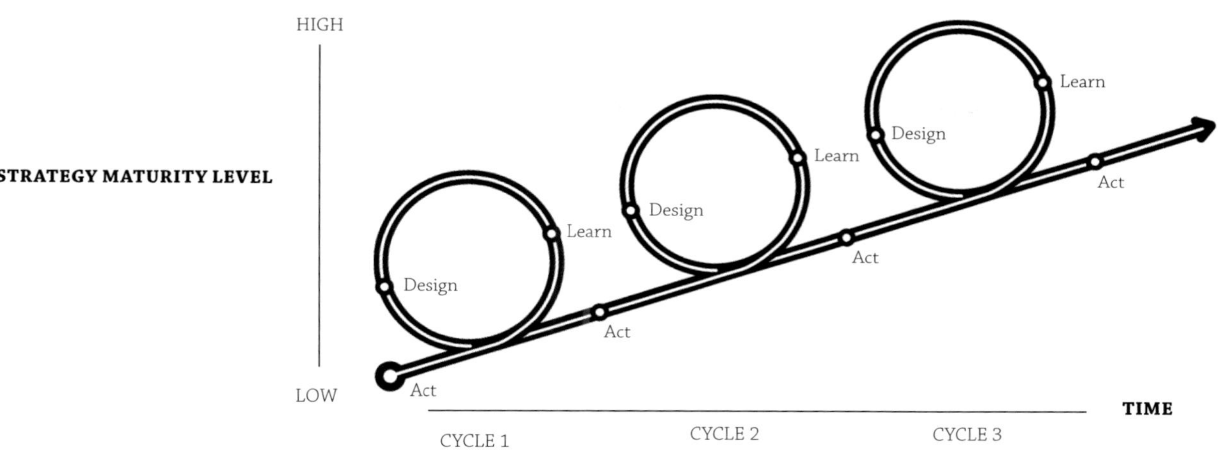

Depending on the nature and size of your business, it is worth thinking about which internal and external stakeholders to involve in these reviews. Some companies like to invite important partners to such sessions. Others might interview customers before a strategy review session and discuss the data collected. Still others invite employees who have a certain level of expertise and firsthand experience for the topics under discussion.

To evaluate what is working well and which elements of your strategy need to be fine-tuned, you also need to track key performance metrics. Start with reviewing your strategy's underlying assumptions. Which are still valid? Which are not? What are the consequences? Check current performance against your objectives. Are you on track to achieving these? No? What does this mean for your strategy? Remember, crafting your strategy is about reflecting and learning about what already works well, and what needs to be changed to increase performance.

WHERE TO LOOK FOR INSIGHTS

Such strategy review sessions might require you to start collecting data well in advance. Talk to customers, partners, and your employees to get a good feeling for how well the strategy is working.

Our research and experience have shown that crafting successful offerings, business models, and revenue models is enabled by a deep understanding of customers and their experiences, as we have outlined in Chapter 2. When fine-tuning your strategy, you will need to go back to them and keep on checking whether your strategy is creating value for them, your partners, and your company along the way.

INSPIRATION

PROSIEBENSAT.1 CRAFTING THE NEW BUSINESS

The following Inspiration case history illustrates how ProSiebenSat.1 (Pro7), the German mass media firm, went from validating its initial idea, through evolving and adapting it based on experiences made, before finally scaling it up.

ACTION > To test whether customers were interested in the new offering, business model, and revenue model, ProSiebenSat.1 posted a press release, announcing it would offer the new media-for-revenue-share model. Within a week, it had dozens of requests by companies to be considered as customers.

After the positive reactions to the press release, ProSiebenSat.1 started to evaluate the various companies that had shown interest in the new business model, drew up the details of the media-for-revenue-share contracts, and approached the first customers to executed deals with them.

LEARNING > After initial trials, it learned that a share of revenues was not enough to maintain its margins, so it further expanded the model to include the "minimum guarantee," a fixed amount for media services, augmented by a share of revenues, calculated based on the positive effects on sales through TV advertising.

It also learned that not every product and type of company is suitable for its model. One of the early customers, for example, was a local soft drinks producer. Although the business looked promising, ProSiebenSat.1 found out (unfortunately, only after having run the TV advertisements) that the drinks were not listed with major supermarkets and the producer did not have the capacity to fulfill demand. It also ran into discussions with companies about how much of the revenues were generated as a result of the TV media. In the case of physical in-store sales, ProSiebenSat.1 had no means to track sales.

DESIGN > As a result of these and similar experiences, ProSiebenSat.1 decided to focus on e-commerce, where additional revenues triggered by advertising could be tracked more easily, and it focused on companies and products with no inventory risk.

ACTION/LEARNING > After having successfully executed the new business model and seeing an increasing volume of business, ProSiebenSat.1 realized that, especially in the case of working with start-ups, other investors were not too pleased with the cash drain that the ProSiebenSat.1 model entailed. It saw that equity participations would bring potentially higher financial returns in the case of a successful initial public offering or sale of the company.

DESIGN > This insight led to the design of a media-for-equity offering as part of the business model.

ACTION/LEARNING > Yet, executing this model brought to the surface certain financial and legal challenges for the media group, which had to be accounted for and discussed at length with the media group's executive management.

DESIGN > As a response to increasing volume of the business, ProSiebenSat.1 designed an organizational structure and processes in which core activities specific to the new business model were consolidated in a dedicated company, while supporting activities (e.g., legal, finance, accounting, controlling, tax, and production) were carried out by the respective departments of the parent company. Two teams were created to manage and execute the core activities of the new business model, and people to staff these teams were hired.

- The **investment team**, consisting of investment bankers having been hired to bring in the necessary skills and capabilities to identify and evaluate the potential of investment targets.
- The **operations team**, which was in charge of managing all processes from due diligence to media planning.

ACTION > The newly established company was successful in helping to create strong brand awareness and an increase in sales for several companies, considered success cases of the new business model (e.g., Zalando, an online shoe retailer, or Tom Tailor, a German casual wear clothing brand with retail shops in 21 countries and a large online presence).

LEARNING > The success of the business model led ProSiebenSat.1 to use it more strategically to invest with media performance in promising markets, choosing the start-ups, products, and services to invest in, while shifting the majority of the reward to equity proceeds or exit.

INSPIRATION

FAHRENHEIT 212 INNOVATING A NEW BUSINESS MODEL

The following case history illustrates how Fahrenheit 212 went from validating its initial idea, through evolving and adapting it based on experiences made, before finally scaling it up.

ACTION > To test whether customers were interested in Fahrenheit 212's new offering, business model, and revenue model, the company contacted potential customers and shared its new strategy with them. One of the customers was so convinced that he bought all of Fahrenheit 212's capacity for a period of six months.

LEARNING > Yet, despite the "amazing customer propositions" created, the "needle on the hit rate did not move at all,"[1] according to the president of Fahrenheit 212. In the eyes of clients, Fahrenheit 212 was extremely successful, but it was not successful by its own measures.

Fahrenheit 212 created a vast amount of ideas, which it saw never being taken to market or not being as successful as it had hoped. It was not enough to impress clients with innovative ideas; the company wanted to create new products, services, and businesses that succeeded in the market and helped clients to achieve considerable growth. "The model was not wrong; it was just not enough to overcome innovation failure," said Fahrenheit 212's president.

The company stepped back and tried to understand the root causes of innovation failure once again. On one side, Fahrenheit 212 saw the world of management consultancies defining growth strategies, but not turning these high-level strategies into tangible offerings that consumers would buy. There was a considerable amount of commercial insight, but no creativity.

On the other side, there were the design companies, which showcased high creativity but lacked the commercial rigor of the management consultancies. These companies tended to work exclusively with a consumer focus: finding consumer problems and designing products and services to solve them, without commercial acumen and orientation toward company strategy.

Fahrenheit 212 realized that each of these two approaches had the piece that the other was missing. Creative and innovative ideas for new products, services, and businesses needed to be aligned with business requirements in terms of financials, (e.g., ROI or margin requirements), operational realities (e.g., whether the company can manufacture it), technology available, and the overall strategic directions of the company.

DESIGN > Hence, the solution to overcome innovation failure was not to develop different or better ideas, as Fahrenheit 212 had tried to do, but to combine creativity (solve for the consumer) with commercial acumen (solve for the business).

To be able to overcome innovation failure and deliver the new approach, Fahrenheit 212 built a new type of practice, labeled "Money & Magic," that was deliberately structured to address the needs of the consumer, as well as the needs of its business clients.

First, the necessary activities to fulfill the commercial promise needed to be developed. Activities focusing on commercial acumen (e.g., market share calculations, financial benchmarking, calculating potential revenues, manufacturing and distribution costs, etc.) had to be integrated with the already existing creative ones.

Second, instead of relying on the traditional sequential process separating execution from ideas, Fahrenheit 212 created a process and working model where both teams, the ideas development team and the commercial strategy team, collaborated from day one on a project.

ACTION > To be able to carry out the new activities, new skills and capabilities were brought into the company by hiring business analysts with the needed experience and background in finance and business, often with an MBA.

A new organizational structure was created to accommodate the ideas and commercial strategy teams. The teams were led by two heads on the same hierarchical level. The president still considers this a vital point in order to establish the right mindset and DNA of the two being equally important to the success of innovation.

Compared to traditional management consultants, teams were no longer co-located to the client's office, but worked in Fahrenheit 212's office, and they were not assigned to a single project, but involved in multiple projects at the same time.

LEARNING > While the new "Money & Magic" business model worked well in terms of creating innovative products and services for clients, Fahrenheit 212 realized that the initial idea of being paid a success fee of 2 percent of the first three years' sales of the product created challenges for its own cash flow. Not only was the timing not optimal, but also there were many crucial steps to making the idea a success in the market that it did not control once the idea had been sold to the client.

Fahrenheit 212 asked itself which steps in the innovation process were the most crucial ones for its clients, and learned that every client had some form of stage gate model, with a number of hurdles and gates to pass, which it used to structure innovation projects and make investment decisions. Innovation managers typically considered it a huge success if an idea was approved to pass a gate and move on to the next stage.

DESIGN > Thus, instead of tying its entire variable fee to the commercial success in the market, Fahrenheit 212 aligned its success fee to the achievement of certain milestones in the client's stage gate process, getting paid each time its ideas passed a gate and overcame a major hurdle inside the company.

ACTION > The model was proposed to new clients to receive their feedback.

Today, Fahrenheit 212 aligns up to two-thirds of its revenues to the achievement of these internal milestones and the commercial success of a product in the market, while always staying flexible and adapting to the client's internal processes.

THE SOLUTION TO OVERCOME INNOVATION FAILURE WAS TO COMBINE CREATIVITY WITH COMMERCIAL ACUMEN.

—**MARK PAYNE** *President and Founder, Fahrenheit 212*

START, STOP, CHANGE, CONTINUE

Use the growth team's insights and observations to develop action steps

OVERVIEW

Teamwork can help to align on the most important activities to start doing, stop doing, change, or continue.

TIME NEEDED

45–60 minutes

MATERIALS

Markers, sticky notes, whiteboard, Opportunity and Strategy Reports

STEPS

1. Print out a copy of your Opportunity and Strategy Reports. By now, you should also have results from whatever form of evaluation metrics you established to measure your strategy. Have participants review all the information and consider how well the strategy is working in practice.
2. Create four columns and write Start, Stop, Change, and Continue as headers.
3. Have every participant reflect on the success of the growth business strategy and give everyone plenty of quiet time to individually write down as many ideas as possible about each of the categories, with one idea per sticky note. Participants will reflect upon:
   - What they should start doing to improve the strategy.
   - What should be stopped that is not effective or working.
   - What aspects of the your strategy should be changed or evolved.
   - What's working well that should be continued.
4. After all the ideas have been posted, map them into categories, and then have a discussion to review the ideas. Gain alignment, and then prioritize the ideas that will have the greatest impact on evolving the growth strategy. Take photos of all whiteboards.

Additional activity resources and templates can be found at www.theartofopportunity.net.

SOURCE: *The Art of Opportunity* authors; Inspiration from *Gamestorming*

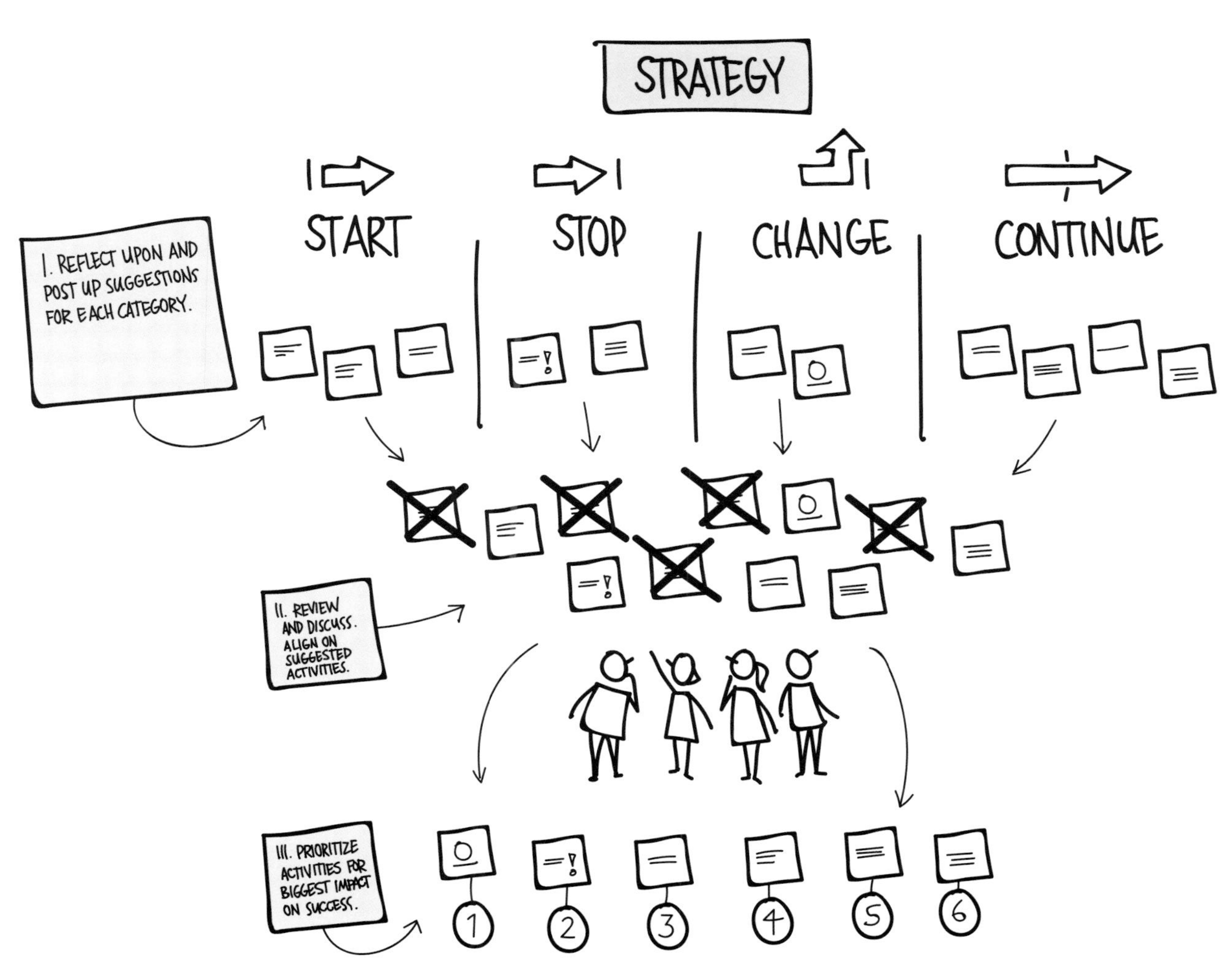
STRATEGY
START
STOP
CHANGE
CONTINUE
I. REFLECT UPON AND POST UP SUGGESTIONS FOR EACH CATEGORY.
II. REVIEW AND DISCUSS. ALIGN ON SUGGESTED ACTIVITIES.
III. PRIORITIZE ACTIVITIES FOR BIGGEST IMPACT ON SUCCESS.
1
2
3
4
5
6

DESIGN NEXT STEPS

Leverage the output of your Start, Stop, Change, and Continue activity session to build an action plan.

OVERVIEW

Use visualized planning to build an action plan for the initiative that will be undertaken by multidisciplinary teams to advance your new growth business.

TIME NEEDED

45–60 minutes

MATERIALS

Markers, sticky notes, whiteboard, and output of the Start, Stop, Change, and Continue activity session

STEPS

1. Using the output from your Start, Stop, Change, and Continue activity session, have a team discussion to craft these into initiatives. Carefully think about the set of skills, knowledge, and resources needed to accomplish each one.
2. Create a grid on the whiteboard, listing each initiative in rows down the first column. Then across the top, in the header, place the name of each functional department that will be a part of the multidisciplinary team to work on an initiative.
3. Choose the team members needed for each initiative and then name a team leader for it. Assign additional roles and responsibilities on each team as needed.
4. Have a group discussion to review all the assignments and plans. Take photos of all whiteboards.

Additional activity resources and templates can be found at www.theartofopportunity.net.

SOURCE: *The Art of Opportunity* authors

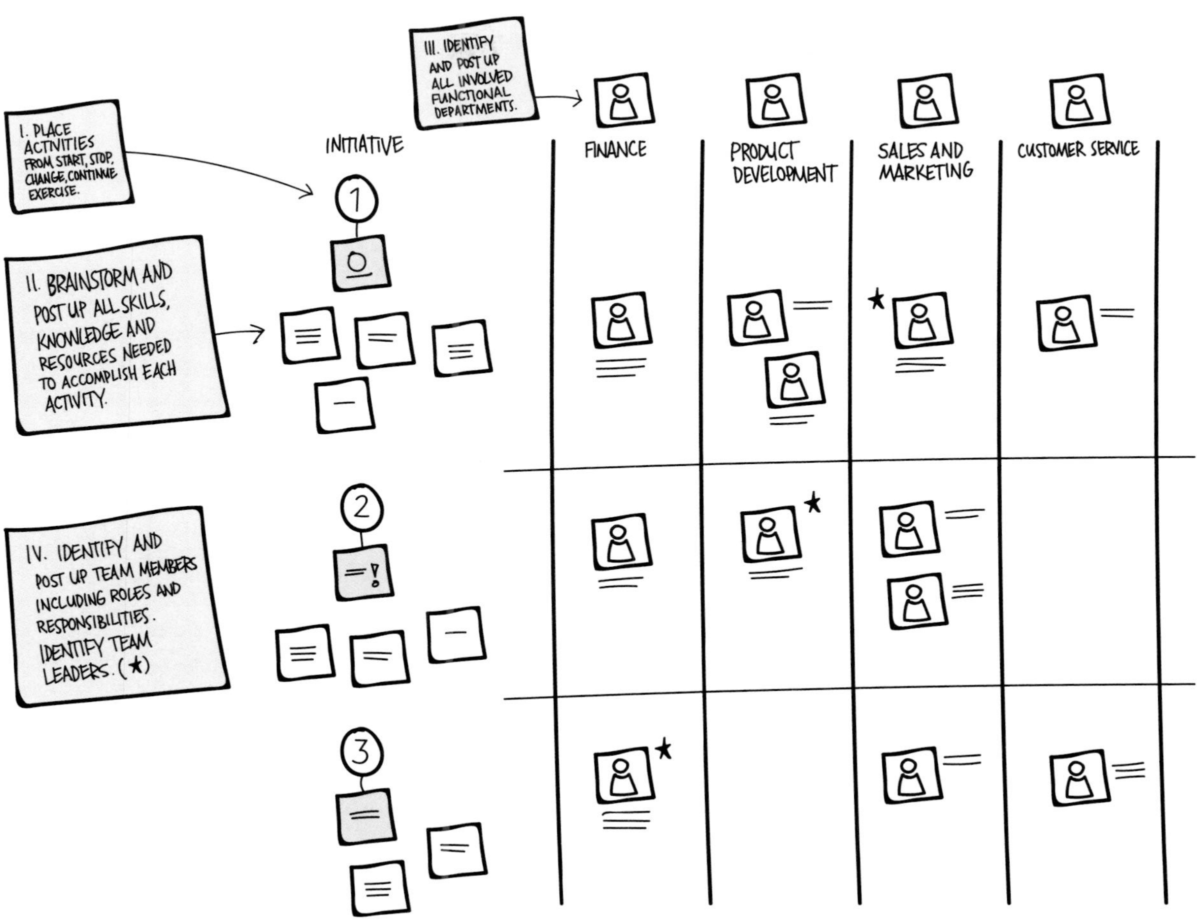
I. PLACE ACTIVITIES FROM START, STOP, CHANGE, CONTINUE EXERCISE.
II. BRAINSTORM AND POST UP ALL SKILLS, KNOWLEDGE AND RESOURCES NEEDED TO ACCOMPLISH EACH ACTIVITY.
III. IDENTIFY AND POST UP ALL INVOLVED FUNCTIONAL DEPARTMENTS.
IV. IDENTIFY AND POST UP TEAM MEMBERS INCLUDING ROLES AND RESPONSIBILITIES. IDENTIFY TEAM LEADERS. (★)
INITIATIVE
FINANCE
PRODUCT DEVELOPMENT
SALES AND MARKETING
CUSTOMER SERVICE
1
2
3

THE EVOLUTION PHASE

VISUALIZE YOUR GROWTH PLAN

Your strategy has been launched and carefully evaluated for improvements that can be made to further evolve your new growth business. Take the results from the Design Next Steps activity.

You will want to illustrate the path you will take to advance your strategy's evolution. Take a holistic perspective, by assigning diverse teams representing key organizational functions and stakeholders to strategy implementation initiatives. Create a clear plan that outlines how the strategy plan is implemented, what happens when, who is responsible, and how success is measured.

VISUALIZE YOUR GROWTH PLAN

Create your plan for building your growth business based on the initiatives you have identified.

OVERVIEW

Illustrate the path you will take to advance the key initiatives of your strategy. Create a clear plan that outlines how the strategy is implemented, what happens when, who is responsible, and how success is measured.

TIME NEEDED

1–2 days

MATERIALS

Markers, sticky notes, and whiteboard, and output of the Design Next Steps activity session

STEPS

1. Bring the output from the Design Next Steps activity session.
2. Create a timeline grid on the whiteboard. Put your timeframes into the headers of the columns. We suggest using calendar quarters for planning. If useful, for the near term work, you can divide the first quarter into 3 months, as most teams spend significantly more focus on the near term activities. To the left of your timeline, create a column for your initiatives that were created in Design Next Steps activity session. To the left of each initiative, have three sub-rows, labeled: Cost and Challenges; Dependencies and Risks; and Measurement and Reporting.
3. Have a group discussion to prioritize growth plan initiatives. Once settled, write the prioritized list in the Initiatives column.
4. Break into teams of two or three and assign each team one or two initiatives. Each group ideates on the consideration topics for each initiative:
   - The cost and operational challenges
   - The dependencies and risks for each initiative
   - How the each initiative is measured and reported
5. Once done, have each team post their initiative consideration ideas on the whiteboard in the corresponding column and row.
6. Each team will present their work, then have a group discussion to review and adjust points in each column category. Then have a team discussion to develop the starting point and ending point of each initiative. Post the agreed-upon timelines for each initiative to the calendar part of your grid.
7. Take photos of all whiteboards and compile the plan into a single sharable document to gather feedback and make regular adjustments to your strategy and execution.

Additional activity resources and templates can be found at www.theartofopportunity.net.

SOURCE: *The Art of Opportunity* authors

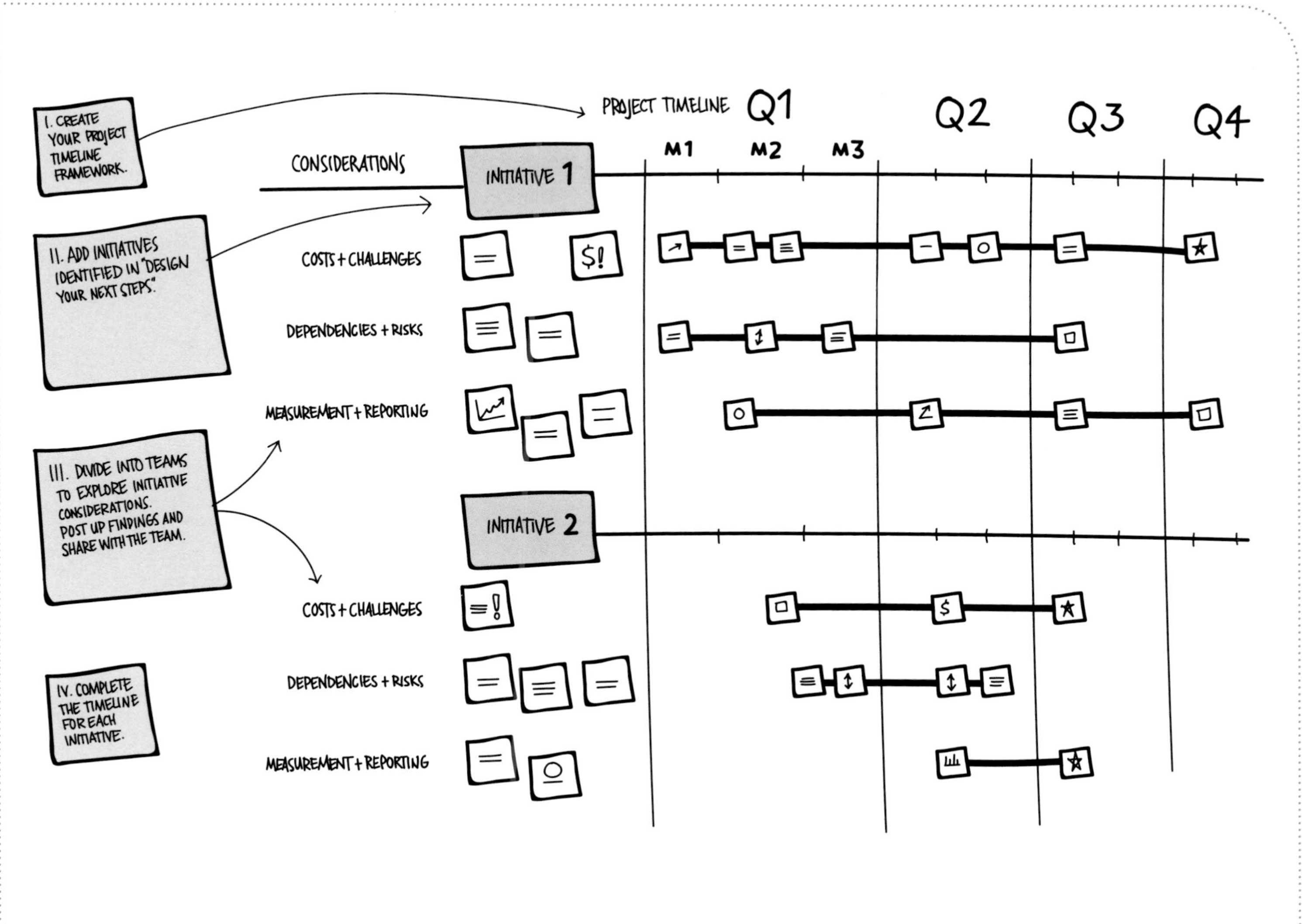
I. CREATE YOUR PROJECT TIMELINE FRAMEWORK.
II. ADD INITIATIVES IDENTIFIED IN "DESIGN YOUR NEXT STEPS."
III. DIVIDE INTO TEAMS TO EXPLORE INITIATIVE CONSIDERATIONS. POST UP FINDINGS AND SHARE WITH THE TEAM.
IV. COMPLETE THE TIMELINE FOR EACH INITIATIVE.
PROJECT TIMELINE
Q1
Q2
Q3
Q4
M1
M2
M3
CONSIDERATIONS
INITIATIVE 1
COSTS + CHALLENGES
DEPENDENCIES + RISKS
MEASUREMENT + REPORTING
INITIATIVE 2
COSTS + CHALLENGES
DEPENDENCIES + RISKS
MEASUREMENT + REPORTING

THE DIFFUSION PHASE: SCALING UP YOUR BUSINESS

ONCE YOUR STRATEGY has reached a certain level of maturity and it is working sufficiently well, it is time to switch from crafting and designing to scaling up your business and rolling it out on a larger scale. Scaling up your business means fully deploying your strategy, acquiring more clients, expanding into new regions, hiring more people, and dedicating more assets and resources. You may also need to create a new organizational entity dedicated to the business.

As the focus of your activities switches, so do the challenges you will encounter. Whereas in the early phases of crafting your strategy you will encounter design challenges, in this phase challenges will tilt toward being of the operational kind.

THINK BIG,

START SMALL,

SCALE INTELLIGENTLY.

—**DERYCK J VAN RENSBURG** *President Coca-Cola, Global Ventures*[2]

THE DIFFUSION PHASE

OPERATIONAL CHALLENGES

Operational challenges come in three modes:

1. Creating awareness and understanding of your new strategy among new and existing employees.
2. Generating commitment and support for the new strategy.
3. Producing action in the form of behavior aligned with the new strategy.

In this stage you no longer design your strategy as such, but take the actions to achieve these three objectives.

Our research has shown that, in the diffusion phase, the main tactics applied by organizations revolve around establishing deployment operations that use top-down direction, and at the same time involve associates working from the bottom up. And your work is not done, as diffusion requires continued attention and evolving. Consider a formal change management program to ensure success.

Consequently, these organizations and their leaders:

- Establish meeting and communication structures that allow providing top-down guidance, while allowing feedback on the new strategy and operational issues to flow from associates to management.
- Adapt organizational structures and processes as needed to make the new strategy work. This includes hiring and also replacing talent quickly as needed to keep up the momentum.
- Establish rigorous project management methodologies to track and follow up on the implementation and success of the new strategy.

CHALLENGES TO LAUNCHING NEW GROWTH VENTURES

	DESIGN-RELATED CHALLENGES	IMPLEMENTATION/OPERATION-RELATED CHALLENGES
COGNITIVE	· Designing a functional strategy · Lack of knowledge which strategy will be successful	· Creating awareness and understanding for the new strategy and business
EMOTIONAL	· Uncertainty and inexperience with the new strategy and how to design it creates discomfort	· Creating commitment and support for the new strategy and business
BEHAVIORAL	· How to design the new strategy, where to look for insights, how to organize design activities · Lack of design experience	· Engaging in implementation and operational activities, acting according to the new strategy

THE DIFFUSION PHASE

OPERATIONAL CHALLENGES

Typical activities include regular top management workshops to review the strategy. Klinik Hirslanden's management, for example, met every six months for major strategy workshops to discuss the new strategy, how implementation was progressing, and which challenges had emerged and needed to be addressed.

To meet your operational challenges, here are some activities you can invoke during your scaling-up effort that will give you a greater chance of success.

ESTABLISH CLEAR ROLES AND RESPONSIBILITIES

It is important to identify an executive sponsor of your initiative. Some organizations create a guiding governance group to oversee the effort. Your sponsor or governing group is not part of the day-to-day activities of the scaling-up effort. The sponsor or governing group's role is managing the strategy, asking tough questions, governing the process, and making key decisions.

Additionally, name a leadership team that owns the scaling-up implementation. You will identify a few internal senior leaders who work well together and are masters of collaboration to be responsible for meeting tactical challenges and managing day-to-day adjustments in operations.

BUILD AN ORGANIZATIONAL CULTURE OF SHARING, LISTENING, COLLABORATION, EXPERIMENTATION, AND SUPPORT

This means having activities, technology, and systems to build community and empower employees to make decisions, take risks, and openly share learning and results. Additionally, part of your sharing, listening, and collaborating is communicating frequently and thoroughly. You will need dedicated resources and leadership for messaging and building engagement with customers, employees, vendors, and other stakeholders.

TAKE A HOLISTIC, SYSTEMS APPROACH TO PLANNING AND EXECUTING THE SCALING-UP PROCESS

This means involving all stakeholders, including all internal employee groups, as well as external groups of customers, vendors, and suppliers for design, review, and feedback of the important aspects of the strategy and business implementation.

BUILD A PLAN AND EMPLOY A PROCESS OF REGULAR AND RIGOROUS REVIEW AND ADJUSTMENT

Details matter, and the process of scaling ebbs and flows in unpredictable patterns. Attention to the small details will avoid large problems down the line. You will want to build a comprehensive plan that addresses every core organizational topic. Scaling up touches every aspect of your business.

Create a clear vision and plan for addressing the following topics:

- Leadership and governance.
- Finance, accounting, and risk management.
- Operations: purchasing, sourcing, manufacturing, and logistics.
- Customer management: sales, business development, marketing, public relations, thought leadership, and partnerships.
- Legal, regulatory, safety, community relations, and governmental affairs.
- Infrastructure and organizational activities: information technology, business processes and administration, data collection and analysis, talent and human development, knowledge sharing, training, and support.

CREATE A CLEAR AND SIMPLE WAY TO TRACK AND MEASURE YOUR PROGRESS

This should include visual dashboards and road maps, so everyone knows where you are going and how you are progressing. The input should be conducted in a systematic means of gathering qualitative (voice of the customer and employee interviews) and quantitative data points. Your leadership team can use insight to make adjustments and evolve your strategy.

MASTERING THE ART: BUSINESS DESIGN THINKING

WE CAN'T SOLVE PROBLEMS BY USING THE SAME KIND OF THINKING WE USED WHEN WE CREATED THEM.

—**ALBERT EINSTEIN** *Theoretical Physicist and Violinist*

Designing and building a new growth business is inherently challenging. The process presented in *The Art of Opportunity* will help you overcome some of these challenges by executing specific steps and activities. But as the master of any craft knows, excellence is not just knowing what to do, but also understanding how to do it. Business design thinking offers a set of underlying principles that, when combined with the strategic innovation process, will increase the speed, efficiency, and quality of your outcome.

WHAT IS BUSINESS DESIGN THINKING?

BROADLY DEFINED, business design thinking is a method-based approach for solving business problems and creating business solutions. It is a way of working based on the idea of identifying and articulating clear goals that are achieved by focusing on audience needs (customer, user, stakeholder, etc.). Delivering on these goals involves engaging multidisciplinary teams with visual thinking methodologies to explore and formulate strategy. During the process, continuous iteration (learning by doing) enables the team to rapidly test ideas and make adjustments based upon their learning. Business design thinking works well with strategic innovation because it opens new channels to creativity, actively engages participants and stakeholders, builds clarity and consensus, and accelerates the speed to market.

FOUNDATION OF BUSINESS DESIGN THINKING

Business design thinking is not a new idea. It has developed over decades through the contributions of inventive thinkers, whose work includes:

- Horst Rittel's pioneering work articulating the relationship between science and design, as early as 1958.
- Lateral thinking, first coined by Edward de Bono in 1967.
- Herbert A. Simon's 1969 book, *The Sciences of the Artificial.*
- Don Koberg and Jim Bagnall's 1971 book, *The Universal Traveler.*
- Robert McKim's 1973 book, *Experiences in Visual Thinking.*
- Ned Herrmann's books, *The Creative Brain* and *The Whole Brain® Business Book.*
- Bryan Lawson's 1979 empirical study, *How Designers Think*, and subsequent books on design.
- Nigel Cross's seminal journal article, "Designerly Ways of Knowing," in 1982.
- Edward Tufte's 1983 book, *The Visual Display of Quantitative Information.*
- Donald Norman's 1986 book, *User Centered System Design.*
- Peter Rowe's 1987 book, *Design Thinking.*
- The work of David M. Kelley, who founded IDEO in 1991.
- Richard Buchanan's 1992 article, "Wicked Problems in Design Thinking."
- Richard Buchanan and Victor Margolin's 1995 book, *Discovering Design.*
- Thomas Lockwood's 2008 book, *Building Design Strategy.*
- Dan Roam's 2009 book, *The Back of the Napkin.*
- *Gamestorming*, the 2010 book by Dave Gray, Sunni Brown, and James Macanufo.
- Alexander Osterwalder and Yves Pigneur's handbook, *Business Model Generation*, in 2010.

PROCESS AND METHODOLOGY

OVER THE YEARS, MANY DESIGN THINKING PIONEERS and practitioners have offered their interpretations of the process. Much of our thinking on this process and its associated principles have been informed by our practical experience with XPLANE and other design thinkers. While the nomenclature differs slightly, there is a general consensus that the process begins with a phase of inquiry and discovery, followed by exploration, design, and conceptualization. The process ends with converging on a final idea or launching the project—depending on the desired goal. The table on the following page offers a comparative outline of some of the processes developed by prominent minds in the design thinking field.

DESIGN THINKING PROCESSES

	DEFINE	RESEARCH	IDEATE	PROTOTYPE	CHOOSE	IMPLEMENT	LEARN
IDEO	Understand	Observation	Visualize	Iterate	Test		
GOOGLE VENTURES	Understand	Define	Diverge	Decide Prototype	Validate		
ALEX OSBORN	Clarify and identify the problem	Research the problem	Formulate challenges; Generate ideas	Combine & evaluate the idea	Draw up an action plan	Do it! (implement the ideas)	
VIIJAY KUMAR	Intent	Context	Users	Insights	Concepts	Solutions	Offerings
	SEEK		**IMAGINE**	**MAKE**	**PLAN**	**BUILD**	
BILL MOGGRIDGE	Constraints	Synthesis	Framing Ideation	Envisioning	Selection	Visualization Prototyping	Evaluation
	INSPIRATION		**IDEATION**		**IMPLEMENTATION**		
PARIS-EST D. SCHOOL	Understand	Observation	POV	Ideate Prototype	Test	Storytelling Pilot	Business Model

PRINCIPLES OF BUSINESS DESIGN THINKING

BUSINESS DESIGN THINKING is not strictly a process, it is an approach. Key principles are intrinsically tied to the processes and embody the practitioner's way of thinking and doing. The five principles of business design thinking are:

1. Keep a human-centered focus
2. Think visually and tell stories
3. Work and co-create collaboratively
4. Evolve through active iteration
5. Maintain a holistic perspective

1. KEEP A HUMAN-CENTERED FOCUS

A deep understanding and empathy for your customer (or noncustomer) is at the heart of business design thinking. By observing, engaging, and immersing yourself in the behavior, needs, and mind-set of your specified audience, you gain critical understanding and insight in order to address their need—and create value.

For many newcomers to business design thinking, this focus on the customer may seem counterintuitive, as they realize that the problem (or opportunity) they are seeking to solve isn't theirs, but rather it belongs to their customer. Keeping a human-centered focus encourages you to discover your customers' needs by "walking a mile in their shoes." This includes observing and potentially participating in their environment to uncover overlooked challenges and other subtle, yet powerful factors that influence their experience. Interaction and mindful

HUMAN-CENTERED

observation of the customer's world can provide valuable insights into their intangible, yet nuanced experience with your offering and in their everyday life. In other words, uncover their stated (and unstated) needs.

The human-centered focus mind-set is characterized by curiosity and the desire to explore the motivations that drive your customers' behavior. But uncovering these insights is not easy. They are seldom found buried in data. Our preconceived notions act as filters to block insights that can be gained through the clear eyes of empathetic observation. This is where visual thinking becomes helpful.

CULTURE'S IMPACT ON GROWTH

Adopting business design thinking can have an impact well beyond the lifespan of your growth efforts. Evidence is beginning to suggest that corporate cultures in which employees are encouraged to maintain a human-centered focus are more likely to experience greater return on investment (ROI).

Google, for instance, promotes their people-centric approach as a key differentiator and primary method for driving innovation. According to Annika Steiber, one of the key principles of the tech firm's management model is employing a people-centric approach by "focusing on the individual and liberating his or her innovative power. This principle is based on a belief that people want to be creative and that a company must provide them with a setting in which they can express their creativity."[1]

This is supported by a Gallup study that found that organizations with an average of 9.3 engaged employees for every actively disengaged employee enjoy 147 percent higher earnings per share when compared to their competition.[2] This is not a bad secondary outcome of adopting business design thinking into your organization.

VISUAL THINKING

2. THINK VISUALLY AND TELL STORIES

While a human-centered focus shows you where to look for opportunities, visual thinking and storytelling help you uncover them and bring them to life. Visual thinking can also help track and make sense of highly complex issues (like the customer journeys and business ecosystems addressed earlier). Additionally, the visualizations created in the process can help you communicate your progress to generate and maintain support for your work within the organization.

Dave Gray describes visual thinking as "a way to organize your thoughts and improve your ability to think and communicate. It's a great way to convey complex or potentially confusing information. It's also about using tools—like pen and paper, index cards, and software tools—to externalize your internal thinking processes, making them more clear, explicit, and actionable."[3]

As a methodology for taking otherwise intangible ideas and making them visible, visual thinking presents a powerful way for groups to analyze and explore problems (or opportunities). The visualization component enables you to share these ideas in a clear, easily understood manner as opposed to in numbers or words. In addition to creating clarity and understanding, bringing stories (or customer journeys) to life visually also generates alignment among stakeholders and can accelerate decision making—essentially speeding up progress on your growth journey.

These stories can also be used to inspire and motivate people to take action. Using visuals to walk others through your progress and share your thinking process (like in the report templates) will help you build support with others in your organization. In fact, researchers at the Wharton School of Business compared visual presentations and purely verbal presentations and found that presenters using visual language were considered more persuasive by their audiences.

Visual thinking can range from simple post-up exercises to more complex drawing activities, neither of which requires significant skill beyond basic facilitation and a willingness to try. While many people share an aversion to drawing in a business setting, a little persistence (again leading by example) and minimal practice (as simple as drawing stick figures) can generate profound results in a very short period of time and set a precedent for long-term best practices among your team members.

3. WORK AND CO-CREATE COLLABORATIVELY

Team-based collaboration is the fuel of any business, whether it's between employees, partners, or customers. It is a driving force for continued efficiency with everyday tasks and a necessity for improving the outcomes of many business activities. Research has found that a great majority of senior corporate executives believe that effective coordination across product, functional, and geographic lines is crucial for growth.[4]

Successful business design thinking favors collaboration and co-creation because an open exchange of ideas in an engaging and highly participatory environment reinforces learning, builds alignment, and generates support among traditionally opposing viewpoints. Additionally, the friction created when people with different points of view and experiences converge generates the kind of creative outcomes that individuals could not have found or done alone.

Cultivating diversity of thought has also been shown to boost innovation and encourage creative problem solving. Why? Different cultures, backgrounds, and personalities actively cultivate unique perspectives. Some people are analytical, whereas others thrive in creative zones. Some are detailed planners, and others love impulsiveness. Intermingling different types of thinkers and perspectives to tackle problems and seek solutions kindles creativity, sparks insight, and improves efficiency.

Further, varying the types of thinkers on your team guards against groupthink, a common issue in teams where the prevailing focus is first and foremost on conformity, often at the expense of good decision making. A diverse group of individuals is inherently more empathetic to new ideas. Additionally, as they work through the problems together, the team is naturally creating support for their ideas. So rather than needing to sell the growth idea across the functions, there are already advocates for the idea within the team and within each department. These advocates are essential to building the support and trust required for successful adoption of the initiative.

ACTIVE ITERATION

4. EVOLVE THROUGH ACTIVE ITERATION

Iteration matters. Business design thinkers relentlessly design and evolve strategies, products, and services through continuous experimentation. They think cyclically, not linearly. They begin with action and learn from the experiences and outcomes at each stage of the growth initiative. These observations fuel further modifications to the strategy or even generate new functions. They are also comfortable with ambiguity and have a willingness to explore.

Embracing active iteration is an essential principle of business design thinking. Alistair Cockburn describes it as "learning by completing."[5] The underlying idea of active iteration is that until you actually build what you are designing, you will not fully understand it.

There are several benefits of applying active iteration:

- **Speed:** Gathering initial requirements and plans is faster with active iteration; you can quickly create premises and assumptions to meet a user's needs, and then develop a prototype to test and compile feedback to quickly home in on the key value to the target user. Further, long-term planning time is reduced, and your effort is focused on discovering and addressing the job to be done for your user.
- **Clarity:** Experimentation works to eliminate ambiguity, as each cycle brings more clarity, and it is less costly and time-consuming. Applying active iteration allows you to break down a complex problem into smaller, testable parts, reducing confusion and more quickly finding answers.
- **Quick improvements:** Each period of reflection in your iterative cycle lets you reassess the key value to be gained by your user, and determine what changes matter and which ones don't pay off. Active iteration cycles allow you to quickly react to market or environmental changes. Longer and more resource-heavy methods move more slowly and are less nimble.

5. MAINTAIN A HOLISTIC PERSPECTIVE

One who has a holistic perspective is mindful of the big picture, seeing the connections among self, group, and the larger ecosystem that bring valuable context to problem solving. As you wrestle with a complex problem, using a systems view illuminates points of interaction and how they are related to other processes. Thoughtful consideration of the connectedness of all elements in an open system puts you in the fortunate position of finding and capitalizing on opportunities. Keeping an open mind, deferring judgment, and being optimistic are characteristics of a holistic systems thinker.

As stated earlier, one of the methods we suggest to find your opportunity is mapping your ecosystem. The objective is to give you a vantage point from which to discover connections and interdependencies among stakeholders, processes, and deliverables. One can apply mapping to systems and processes of any size, from a small group to an entire global business.

Thinking holistically with an integrated systems view, you can identify the patterns within a web of relationships and see the values between the components of your system. You can also begin to identify how individual elements within your system work with and relate to each other. Simply put, viewing the organization as a dynamic, open system with interrelated processes creates opportunities to better understand context and improve how the elements work together.

INSPIRATION

ORACLE INNOVATION THROUGH COLLABORATIVE CO-CREATION

How can you establish and institutionalize a collaborative system of co-creation in your organization?

In 2014, Oracle launched *Mission Red*, an EMEA-owned and focused program, to identify, share, and drive ideas. The goal is to turn these ideas into innovations which would create value for customers, and Oracle's business (including its employees, communities, and ecosystems within which it works).

While many corporate innovation programs seek to encourage innovation, *Mission Red* recognizes the creativity of Oracle's people and provides a tangible process for them to get their ideas recognized, developed, and delivered.

The objectives of Mission Red are to:

- Programmatically drive the recognition and delivery of innovative ideas;
- Allow innovative ideas of different size and complexity to be applied by the teams that need them;
- Increase the speed and visibility of idea sharing, collaboration, and best practice across the business.

Mission Red is built around 3 core drivers. These drivers act as lenses to evaluate specific proposed ideas.

- **Innovate:** Think differently, innovate consistently, test assumptions, and put into practice
- **Collaborate:** Work together, build teams, use the best skills, identify best practice, and make progress
- **Accelerate:** Move quickly, review ideas regularly, measure, and recognize achievement

At the heart of *Mission Red* is the Innovation Methodology, a five-stage process defining how ideas move from an initial concept through a final delivered innovation, which recognizes the diversity of people needed in the innovation process.

1. **Create:** Ideas can happen anywhere and at any time. *Mission Red* encourages all employees to find their inspiration and write down their ideas, wherever and whenever they arise.

2. **Share:** Ideas need to be shared, using the skills and experience of others to refine, validate, and hone them, harnessing the process and specialists across the Oracle global team.
3. **Decide:** Eventually, a decision needs to be made. Management teams, large and small, are empowered to review ideas from any proposer and commit to supporting them with resources to deliver business value.
4. **Apply:** With the visible support of management and designated resources, the project team works to turn the perceived value of an idea into a working innovation, testing its validity, quantifying the value, and turning the idea into a delivered innovation.
5. **Accomplish:** With innovation achieved, Oracle celebrates the success, recognizes the individuals involved throughout the process, and enables the innovation to be shared and re-implemented in other relevant parts of the business.

The *Mission Red* Innovation Methodology is supported by the Innovation Engine, a cloud-based tool supporting the creation and sharing of ideas, the public refining of ideas within a social tool, public project management to engage contributors and experts, and visible recognition of innovation champions.

After 18 months more than 20,000 people engaged in the Innovation Engine, creating more than 1,000 ideas, 100 projects, and 30 completed innovations.

INSPIRATION

EXACTTARGET USING VISUAL THINKING AND STORYTELLING TO FIND OPPORTUNITY

ExactTarget was a developer of enterprise SaaS marketing automation applications, before being acquired by Salesforce in October 2013, which was at the time the largest SaaS acquisition in history.

Part of ExactTarget's legacy of was a strong commitment to its partner community. In 2011, it had over 450 partners worldwide. One of the firm's longstanding challenges was how to get alignment on some of the most important issues and needs facing this critical component of their business model.

ExactTarget hoped to expand its market and continue its run of 11 consecutive years with double-digit growth. To help achieve this goal, VP of Channels, Jim Kreller, and Tom Williams, Director of Partner Programs and Operations, took a bold approach to partner engagement at the firm's annual worldwide partner summit.

Working with the strategic design consultancy, XPLANE, they brought together 300 attendees representing dozens of partners from around the world and put them to work creating solutions to have ExactTarget become an industry-leading partnering organization. ExactTarget and XPLANE designed a crowdsourced approach using participation from all the partner attendees to address these challenges. The partners were segmented into 6 separate rooms; each room had 5 tables each with 8 participants.

Using a four-hour brainstorming activity, each table was given a challenge. These challenges were made up of some of the largest obstacles the firm faced in partnering: for example, "If you have limited budget and limited staff, what would be the highest level of support you would want (e.g. marketing, sales, technical, services) as a partner and how would you prioritize that support?" Other challenges involved pricing, quoting, enablement, Market Development Funds, program entitlements, and monetization models.

The challenges were selected for degree of difficulty and impact on the ExactTarget business. Each table spent time ideating, discussing, and collaborating to create their solution, and then elected a speaker to present their results to the room. The best solutions, as voted on by the partner participants, were then presented to the ExactTarget Partner Advisory Council, which was made up of a team of senior executives and principals at several of the company's most strategic partners.

The process was highly collaborative and built upon an inventive co-creation method. Partners were engaged, animated, enthusiastic, passionate and, in some cases, very emotional.

The outcome for ExactTarget was astounding. The session generated solution blueprints that addressed problems that in many cases had existed without a solution for years. In just 4 hours, a set of very innovative ideas were formulated, validated by partner community consensus, and endorsed by the large, and influential, group of partner advocates.

A NEW WAY OF WORKING

BUSINESS DESIGN THINKING isn't for every company or every employee. To effectively implement business design thinking requires a change in mind-set for the individual and for the firm's culture. Many firms struggle with change and innovation for any number of reasons. Some have a culture more accustomed to maintaining strict adherence to processes. Others may have rigid silos for organizational functions and operational formality. Yet another firm may adhere to strict rules for communication and interaction with customers and stakeholders that alienate employees. In these instances, implementing business design thinking practices will be very difficult.

We believe that anyone can become a business design thinker. However, not everyone will be comfortable with the accompanying ambiguity or demands of co-creation. It is not uncommon for people to feel frustrated, uncomfortable, lost, and sometimes even angry when employing business design thinking. That's because changing how you work isn't something that happens overnight. The better you understand your culture and the capabilities of your team, the more likely you will be to successfully introduce new ways of working. Start small. Introduce one practice at a time to just the members of your growth team. And lead by example. Like any change, it will take time to become habit, but if you work as a team and learn from your experiences, you will begin to see the benefits.

Our research and experience have shown us that the business design thinking way of working empowers employees to use their judgment to make the right decisions, builds trust and respect from their peers, and supports and rewards them for their creativity and initiative. The impact that this internal change has on customers can vary, but it is almost universally positive—ranging from improved product quality to more consistent brand experiences.

INSPIRATION

MICROSOFT TRANSFORMING FOR NEW GROWTH

"We as a company stand for deeply understanding the needs of customers, translating that understanding into products that people love and ultimately into the success our customers have with our products. It's that last part that is our key motivation. The entire Microsoft team is inspired to bring their best ideas and efforts every day to build products people love, and to advance our mission to empower every person and every organization on the planet to achieve more. And we're seeing the impact."[6]

—**Satya Nadella,** *CEO, Microsoft*

When Satya Nadella became CEO of Microsoft in 2014, he faced a daunting challenge. Since its inception in 1975, Microsoft has always sold packaged software and was the king of the tech industry in the 80s and 90s. But when Nadella took the reins, people wondered if Microsoft could restore its reputation for innovation and creativity. Some industry pundits argued the company's best days were behind it. Although Microsoft has through the years continued to make a lot of money—delivering billions in revenue and returning profits to shareholders in the form of dividends—its core business was declining amid a slump in the market for PCs.

The question Nadella and his leadership team faced: how to grow the business once again to become the leader of the tech world. "We had to decide, will Microsoft remain a viable and relevant company in the future?"[7] according to Bob Kelly, Corporate Vice President on Microsoft's Cloud+Enterprise business. Clearly, resorting to the traditional strategic moves it had taken, working to expand existing core business and offerings, would certainly not be sufficient.

OLD SCHOOL MODEL

The firm's traditional strategy integration for its offering, business model, and revenue model was well practiced and had been wildly profitable. It built versions of software programs over 1–2 years, then released directly to consumers, as well as through a network of independent software vendor organizations (ISVs).

However, with the market quickly shifting from packaged software to cloud services, the tech firm didn't pivot quickly enough. Faced with intense pressure to save money, IT departments that historically bought packaged software solutions were evaluating software as a service (SaaS) and cloud-backed software solutions to replace or augment their legacy packaged software solutions.

Although they told the world that cloud computing was the future as early as 2008, Microsoft's flat stock price suggested they had not capitalized on the initial ride of two huge market waves: (1) consumers and businesses' need and desire to use apps and enjoy greater mobility and (2) the desire to perform all activities in the cloud and using SaaS.

TIME FOR A NEW OPPORTUNITY

Something obviously had to change. Microsoft's employees and leaders knew it. "Opportunities still exist. We were not paying attention to the market and did not take full advantage of the new market waves. We didn't completely miss it, but we were clearly late," said Kelly.

We found that Microsoft applied many of the elements of strategic innovation and business design thinking to "turn the ship." The firm crafted a unique strategy to completely transform the company—focus on meeting users' needs in the explosive markets of mobility and cloud computing, venturing into new offerings, changing how it did business and how it worked with its ecosystem of partners, and setting a course to transform its culture. Let's take a look at some of Microsoft's actions.

INSPIRATION

CHANGE OFFERINGS? WELL, ACTUALLY, CHANGE EVERYTHING

While Microsoft began to plan and build cloud-based products and services, such as Azure, as early as 2008, the tipping point was in 2011, when Nadella wrote his now internally famous "cloud first" memo, which defined the new path for the Cloud and Enterprise division and Microsoft. He said not only is cloud important, it is cloud first. That memo kicked off massive investment shifts, organizational and cultural changes, and significant revision of business models. The new world order and focus of Microsoft's strategy is called, "Mobile-First, Cloud-First."

Microsoft is in the midst of its own massive transformation to become the productivity and platform company for the mobile-first and cloud-first world. At its core, the company is moving from selling packaged software to selling services via subscription. It's a journey that could take years, but Microsoft is not only committed but is making significant rapid advances along many fronts.

BUSINESS MODEL

To further enable its transformation, Microsoft had to look at customers and the industry in a new way, partnering and co-developing, instead of simply charging for software in a box. No longer are there fees for annual maintenance and updates. Instead, as part of the shift in its revenue model, Microsoft applies a tiered subscription of monthly fees for subscriptions and customers can sign up without expensive long-term contracts and leave when they choose.

And the distribution model has obviously changed to digital and cloud-based, which also means the price points for everything have changed. Customers can buy in small price doses, from multiple possible sources. The democratization of software is an ongoing trend that forces the customer and seller together in a "bear hug" relationship. Microsoft's transformation is made even more complex than others, because the firm's product lines touch nearly every corner of corporate and personal computing.

A key benefit of the shift in business model was creating more value for customers by removing barriers to consumption, and providing a better customer experience. No longer did customers need to invest in expensive hardware and servers, and pay for server setup and maintenance, nor purchase separate seats of Microsoft Office, Exchange, or SharePoint. The new cloud-based offering made it easier for customers, who no longer had the cost and pain of having to install and configure their own corporate server.

The business model also shifted to partnering with customers, having a mind-set to experiment, shoulder to shoulder, to find a solution to a problem. The customer is now a collaborator, working Microsoft in small, iterative, hypothesis-based engagements. Microsoft has changed its view of business partnerships, from transaction-based to emphasizing relationships. Under Nadella's leadership, partnerships are now seen as gateways to innovation and unlocking mutual value—even if that may be with competitors.

REVENUE MODEL

Microsoft's enterprise sales team altered their revenue model. They had always been trained to sell large volume (6 and 7 figure) deals. It was not uncommon for a salesperson to have a $30M quota. In the new cloud world, the sales strategy had to shift to a "land and expand" sales approach. And the initial deals might be as low as $25K, leaving the sales team crying, "How can I make quota?!" The existing business model didn't allow them to succeed. So the way of viewing and measuring success had to be altered.

INSPIRATION

OFFERING DEVELOPMENT MODEL

Another shift for Microsoft's offering has been in the product development model and the process, operations, and the culture supporting it. Microsoft labels it hypothesis-based engineering, a perfect example of the iterative action business design thinking principle. The firm's engineering groups changed their software development methods to modern devops, through the use of agile methodology, which they dub hypothesis-based engineering, instead of the slow, cumbersome and costly old "waterfall" method. "In a cloud world where delivery is constant, hypothesis-based engineering changes how you develop, design and deliver software. It changes everything—full stop—EVERY single thing about your business. The offering, its development, the customer ecosystem, sales, deal nature, service and support," said Kelly.

Hypothesis-based engineering, which now runs the business, is at the heart of the new development process. The organization's focus is now based on 6 pillars:

1. Transform data center (cloudification—putting Azure as a node on your network)
2. Mobile-izing the end user experience—through the enterprise mobility suite
3. Enabling modern app development (a new approach to building apps using the cloud, with dockets, containers using cloud services)
4. Driving insight from data—what do I do with it? How do I drive my business through data? Data is the new platform—betting the farm on voice search—interacting with data in a new way; the ability to search with your voice.
5. Modern Developer Tooling—moving from a toolset on your desktop to design, build, deploy, and operate in the cloud.
6. "Internet of Things"—by 2020 there will be 70 billion intelligent and connected devices. Harnessing that intelligence requires smart devices and a cloud to manage the data and to drive business insight.

CREATING VALUE THROUGHOUT ITS ECOSYSTEM

Microsoft also reengineered its ecosystem model and all its inherent interactions. Once viewed as protective of its own ecosystem and eager to crush its competition, Microsoft is now viewed as something else to even its staunchest rivals—a friend. In the past year, for example, Microsoft has partnered with Salesforce, Red Hat, Dropbox, and others widely seen as competitors.

The hypothesis-based engineering model has also made Microsoft more agile, working to demonstrate how quickly they can develop, partner, and deploy co-created applications and services. Instead of simply providing software, they are co-designing and iterating with customers and partners, in rapid turn development, bringing customers significantly more value.

Having once been viewed as a lumbering behemoth, Microsoft has worked to become much more agile, what they call running a marathon at a sprinter's pace. According to Kelly, they have had start-ups saying, "I don't know how you got back to me so quickly. You're faster than I am!" Some of that new value creation Kelly attributes to Microsoft's adoption of a "growth mind-set," a philosophy originally pioneered by Stanford psychologist, Carol Dweck, that encourages a culture of learning and risk taking often embraced at startups.

A NEW MIND-SET OF BUSINESS DESIGN THINKING

In addition to strategy, business model, revenue model, and offering development changes, Microsoft is undertaking a massive cultural evolution. The firm has changed the cadence and process of how employees work, as well as their focus and behavior. Staff and leaders are now taking a much "humbler approach" to the market, based on the well-grounded realism on where they are today. The importance of relevance is deeply ingrained in Microsoft now.

Nadella continued a practice started by his predecessor, Steve Ballmer, by setting a weekly meeting cadence for his senior leadership team (SLT). Once a week, for an entire day, the SLT sits down together and reviews a key strategic issue and a specific timely topic. Ground rules: be transparent, honest, and respectful. Each week, someone on the SLT is the meeting "cop" to enforce this behavior. This meeting structure helped break down silos within the company.

To instill inspiration and foster impactful work, each week a direct report opens the meeting with an "awesome moment" describing an inspired behavior or action from someone on their team. The desired outcome is creation of an environment where the celebration of awesome is part of the culture.

The energy and outlook of the company's employees is also changing. Every year they have an employee poll to take the pulse on how employees feel about their work and their job. Formerly, the lowest-rated question was, "I believe Microsoft is making the right choices for long term success." Now the poll question's response is at an all time high (measured every year for last 8 years) and is 17 percentage points above the all time average. The company's objective is to provide value to its employees, so they have a meaningful experience doing their job and feel they are creating impactful work every day. They want everyone to feel impassioned.

An interesting talent innovation at Microsoft is something called "Moneyball," which is a new process for onboarding engineers. We see it as an example of the business design thinking of interactive action. The new employees learn that their work always starts with a hypothesis. They then go out and test it, learn, adjust, and try it out again. This process is a whole new way of thinking. And the firm even takes the new employees "outside," to different regions to observe, learn, and practice Moneyball hypotheses-based methodology in action.

Results? The enterprise mobility suite team has delivered the fastest growing enterprise product Microsoft has ever released.

WHERE IS MICROSOFT NOW?

While the transformation is far from over, the effects of strategic innovation, applied in the firm's offering, its business model, revenue model, and customer experience, along with changes in the operations of its resources and assets, has had a huge impact. Not to be unnoticed is the change in mind-set and way of working, with all employees focused on delivering relevance and value to Microsoft's customers.

At the closing of 2015, in a year in which the Standard & Poor's 500 has been basically flat, Microsoft has been hitting new 52-week highs and is up nearly 20% from when Mr. Nadella took over, largely on the strength of its commercial cloud business.

Nadella and his leadership team are working toward two major milestones, reaching 1 billion Windows 10 devices in FY2018 and achieving a $20 billion annualized revenue run-rate in commercial cloud. Already Microsoft is making significant progress on both fronts. At the same time, the company has undertaken an ambitious plan to invent new categories with HoloLens, the first fully untethered, holographic computer, and Surface, its family of tablets to replace the laptop, receiving rave reviews from developers, consumers, and enterprises adopting the new technologies.

OUR INDUSTRY DOES NOT RESPECT TRADITION —IT ONLY RESPECTS INNOVATION.

—**SATYA NADELLA** *CEO, Microsoft*

THE JOURNEY OF A THOUSAND MILES BEGINS WITH A SINGLE STEP.

—**LAOZI** *Philosopher and Poet*

That's our journey. The *Art of Opportunity* journey has presented a broad set of concepts and has outlined a new process for finding your growth opportunity by moving from traditional strategic planning to new strategic innovation. You've learned to craft your strategy from a customer-focused opportunity and to design your offering, business model, and revenue model. Finally, you found out how to build your new growth business to create value for customers, your firm, and your ecosystem.

These concepts are inherently tied to a new way of working based on the principles of business design thinking, with a rigorous process requiring you to iteratively move back and forth between doing and thinking while steadily evolving your growth strategy and ultimately launching your new business.

It's a journey—one of continual iteration. Get going!

NOTES

1 ARTFUL INNOVATION

1 "Annual Report 2014." ProSiebenSat.1. 2015. http://annual-report2014. prosiebensat1.com
2 To make our point, we are presenting the basics of strategy in a very simple way here. For a more elaborate explanation, we suggest reviewing Bob De Wit and Ron Meyer, Strategy: *Process, Content, Context* (Boston: Cengage Learning, 2010).
3 M. E. Porter, *The Competitive Advantage: Creating and Sustaining Superior Performance* (New York: Free Press, 1985) 2005.
4 Michael Treacy and Fred Wiersema, *The Discipline of Market Leaders* (Addison-Wesley, 1995).
5 Carol S. Dweck, *Mindset: The New Psychology of Success* (New York: Random House, 2007).

2 DISCOVER YOUR NEW GROWTH OPPORTUNITY

1 See, for example: Barry Jaruzelski, Volker Staack, and Brad Goehle, "The Global Innovation 1000: Proven Paths to Innovation Success," *Strategy+Business*, October 28, 2014, www .strategy-business.com/article/00295; M. R. Dixit, S. Sharma, and A. Karna, "Strategic Breakthroughs as Flagpoles of Innovation," *International Journal of Business Innovation and Research*, 2014; A. García-Granero, J. Vega-Jurado, and J. Alegre, "Shaping the Firm's External Search Strategy," *Innovation* 16, no. 3 (2014): 417– 429.
2 To the best of our knowledge, the idea of "noncustomers" was first introduced by W. Chan Kim and Renée Mauborgne, *Blue Ocean Strategy (First Edition): How to Create Uncontested Market Space and Make Competition Irrelevant* (Boston: Harvard Business School Press, 2005).
3 W. Chan Kim and Renée Mauborgne, *Blue Ocean Strategy (Expanded Edition): How to Create Uncontested Market Space and Make Competition Irrelevant* (Boston: Harvard Business School Press, 2015).
4 See, for example, Clayton M. Christensen, Scott Cook, and Taddy Hall, "Marketing Malpractice: The Cause and the Cure," *Harvard Business Review*, December 2005.
5 Clayton Christensen, https://youtu.be/f84LymEs67Y.
6 These resource Spark questions came from Strategic Management Insight (www.strategicmanagementinsight.com).
7 Moore, James F. *The Death of Competition: Leadership and Strategy in the Age of Business Ecosystems*. New York: HarperBusiness, 1996.

3 CRAFT YOUR STRATEGY

1 The four perspectives are based on a rigorous analysis of more than 200 academic and managerial publications on the business model concept. For a detailed explanation, see Marc Sniukas, "The Micro-Foundations of Business Model Innovation as a Dynamic Capability" (doctoral thesis, University of Manchester—Alliance Manchester Business School, 2015), www.sniukas.com/dba.
2 Focusing on the way a company operates and the associated activities is a widely accepted definition of the business model. See, for example, Sniukas op. cit. or C. Zott and R. Amit, "Business Model Design: An Activity System Perspective," *Long Range Planning* 43, nos. 2–3 (2010): 216–226, http://doi.org/10.1016/j.lrp.2009.07.004.
3 This list was inspired by Robert S. Kaplan and David P. Norton's *The Balanced Scorecard: Translating Strategy into Action* (Harvard Business School Press, 1996).

4 LAUNCH YOUR NEW GROWTH BUSINESS

1 Marc Sniukas, "The Micro-Foundations of Business Model Innovation as a Dynamic Capability" (doctoral thesis, University of Manchester—Alliance Manchester Business School, 2015), www.sniukas.com/dba.
2 Deryck J van Rensburg, "Strategic brand venturing: the corporation as entrepreneur," *Journal of Business Strategy*, 2012, Vol. 33 Iss: 3 pp. 4–12)

5 MASTERING THE ART: BUSINESS DESIGN THINKING

1 Steiber, Annika. "How Google Manages Continuous Innovation in a Rapidly Changing World - Strategos - Strategy and Innovation Consulting Firm." Strategos Strategy and Innovation Consulting Firm. March 7, 2014. Accessed January 31, 2016. http://www.strategos.com/google-model-managing-continuous-innovation-rapidly-changing-world/. For more on Google's innovation culture see also Annika Steiber, *The Google Model: Managing Continuous Innovation in a Rapidly Changing World, Management for Professionals* (New York: Springer, 2014).

2 2013 Gallup Report, "State of the American Workplace."

3 Dave Gray, http://xplaner.com/visual-thinking-school/

4 See, for example, Robert L. Cross, Roger D. Martin, and Leigh M. Weiss, "Mapping the Value of Employee Collaboration," McKinsey & Co., August 2006, or Anesa "Nes" Diaz-Uda, Carmen Medina, and Beth Schill, "Diversity of Thought and the Future of the Workforce," Deloitte University Press, 2013.

5 Alistair Cockburn, "Incremental versus Iterative Development," http://alistair.cockburn.us/Incremental+versus+iterative+development

6 Microsoft 2015 Annual Report.

7 Kelly, Bob. "Interview with Bob Kelly." Telephone interview by author. December 31, 2015.

ABOUT THE AUTHORS

MARC SNIUKAS partners with leadership teams and their organizations to discover opportunities for new growth, develop breakthrough strategies and innovative business models to seize those opportunities, and transform their organization to execute the new growth strategy.

He has worked with major corporations across a wide range of industries in Europe, the United States, Canada, Latin America, Russia, Turkey, Saudi Arabia, the United Arab Emirates, Singapore, China, and South Africa.

Besides his corporate work, Sniukas is currently an adjunct professor at the Monterrey Center for Higher Learning of Design (CEDIM), Santa Catarina, Mexico, where he teaches Innovation and Business Model Innovation within the Master of Business Innovation program.

Sniukas has also run courses and given guest lectures at leading business schools, including Stanford's Graduate School of Business, the Boston University School of Management, and the European Business School London, among others.

As an entrepreneur, Sniukas has worked in the music industry and was the founder and managing director of a training organization developing and delivering executive programs on strategy, leadership, general management, and business administration for companies worldwide. The company was successfully sold.

He is currently a cofounder of the Business Model Gallery at businessmodelgallery.com, the world's largest business model database, and a partner at Doujak Corporate Development.

Sniukas holds a doctor of business administration (DBA) degree from Alliance Manchester Business School and a master's degree in business, economics, and social sciences from the Vienna University of Economics and Business, and he studied training and organizational development at Salzburg University Business School. He is also a certified Blue Ocean Strategy Practitioner and has had extensive training in group dynamics and process facilitation.

Additional Credits

- Long-term research (10+ years) on the topic of strategic innovation.
- Doctoral research at Alliance Manchester Business School.
- Recognized expert: "Marc is a global expert on strategic innovation." —Wayne Simmons and Keary Crawford, authors of GrowthThinking.
- Advisory, consulting, and training at Fortune 500 companies on a global scale.

PARKER LEE, a veteran of the technology, entertainment, and sports marketing industries, currently serves as president of Compass52. He approaches his work as a business design impresario, applying his international experience designing and directing large-scale interdisciplinary projects for corporate clients and running and managing companies.

Most recently, Lee was president and executive vice president of business development at XPLANE and was responsible for sales and marketing efforts at the company, which enjoyed significant annual growth and delivery of innovative design thinking engagements for global clients during his eight-year tenure.

Lee has been actively designing organizations for better performance since the 1970s. During the dot-com era, he acted as vice president of business development for four pre-IPO technology companies. He also led sales and licensing efforts for FeatureCast, a producer and syndicator of online digital content, and served as general manager and vice president of sales for eCal, a calendaring technology platform.

Among his career stints, Lee pioneered the use of social media when he was responsible for the development and implementation of online marketing and communications for the California State Democratic Party during the 2004 election.

Prior to breaking into the technology field, Lee enjoyed a varied career in the entertainment and sports industries. Notably, in the 1980s he acted as director of entertainment and special events for Caesars Palace Las Vegas Hotel and Casino, and as agent for cycling star Greg LeMond.

Lee has worked with dozens of global clients, including Cisco, Microsoft, BP, Shell, Intel, Nike, InterContinental Hotels Group (IHG), Autodesk, PayPal, Bank of America, *The Economist*, UNICEF, the World Bank, Genentech, AT&T, Michelin, and many more. He has led many workshops, facilitated events, and provided consulting services for the design consultancy. He has presented at numerous industry events, most recently at the Association of Change Management Professionals annual conferences and the Apple CEO Conference on the Future of Media.

Lee holds a bachelor's degree in organization development and long-range planning from the University of California, Davis. He is currently co-chairman of the Business Committee for the Arts in Portland, Oregon, and past board chairman of the Optical Storage Technology Association (OSTA) and the San Francisco nonprofit organizations Theatre Bay Area and Dance Through Time.

MATT MORASKY is an award-winning creative director, business design consultant, and visual thinker who comfortably straddles the business and creative worlds. Morasky combines 15 years of design thinking experience with facilitation and visual problem-solving practices to bring clarity to complex challenges for global clients including: American Express, British Telecom, Coca-Cola, Elsevier, Intel, InterContinental Hotels Group, Microsoft, The North Face, Red Bull, Swisscom, and Vans.

Morasky currently works at XPLANE, a business design consultancy based in Portland, Oregon and Amsterdam, Netherlands. In 2010 he helped launch XPLANE's Amsterdam studio where he also established the Amsterdam Visual Thinking School (VTS)—designing, promoting, and delivering monthly workshops for the creative and business communities. As an active advocate of visual thinking he most recently developed and launched the Designer Training Program, a multi-week, open-source course that introduces visual thinking (and design thinking) methodologies to senior-level designers and problem solvers.

A natural teacher, Morasky has also taught and regularly guest lectures in many design schools in the U.S. Northwest. He has conducted workshops on visual thinking throughout Europe and the United States, as well as spoken on the topic for audiences at the University of Michigan Ross School of Business, the Frontiers of Interaction conference in Rome, and the Oregon Market Research Association.

Morasky holds a degree in political science from Boston College and an associate's degree in design and illustration.

He lives in Portland, Oregon with his son and dog.

INDEX

A

B

P

R

S